THE STEPHEN BECHTEL FUND
IMPRINT IN ECOLOGY AND THE ENVIRONMENT

The Stephen Bechtel Fund has established this imprint to promote understanding and conservation of our natural environment.

The publisher gratefully acknowledges the generous contribution to this book provided by the Stephen Bechtel Fund.

MIRROR LAKE

FRESHWATER ECOLOGY SERIES
WWW.UCPRESS.EDU/GO/FWE

Editor in Chief: F. Richard Hauer (Flathead Lake Biological Station, University of Montana)

Editorial Board

Emily S. Bernhardt (Department of Biology, Duke University)

Stuart E. Bunn (Australian Rivers Institute, Griffith University, Australia)

Clifford N. Dahm (Department of Biology, University of New Mexico)

Kurt D. Fausch (Department of Fishery and Wildlife Biology, Colorado State University)

Anne E. Hershey (Biology Department, University of North Carolina, Greensboro)

Peter R. Leavitt (Department of Biology, University of Regina, Canada)

Mary E. Power (Department of Integrative Biology, University of California, Berkeley)

R. Jan Stevenson (Department of Zoology, Michigan State University)

University of California Press Editor: Charles R. Crumly

MIRROR LAKE

Interactions among Air, Land, and Water

Edited by
Thomas C. Winter and Gene E. Likens

UNIVERSITY OF CALIFORNIA PRESS   Berkeley   Los Angeles   London

University of California Press, one of the most
distinguished university presses in the United States,
enriches lives around the world by advancing
scholarship in the humanities, social sciences, and
natural sciences. Its activities are supported by
the UC Press Foundation and by philanthropic
contributions from individuals and institutions. For
more information, visit www.ucpress.edu.

Freshwater Ecology Series, Volume 2
For online version, see www.ucpress.edu.

University of California Press
Berkeley and Los Angeles, California

University of California Press, Ltd.
London, England

© 2009 by the Regents of the University of California

Library of Congress Cataloging-in-Publication Data

Mirror Lake: interactions among air, land, and water /
edited by Thomas C. Winter and Gene E. Likens
    p.   cm.
  Includes bibliographical references and index.
  ISBN 978-0-520-26119-8 (cloth : alk. paper)—
ISBN 978-0-520-94449-7 (online edition)
   1. Lake ecology—New Hampshire—Mirror
Lake (Grafton County). 2. Limnology—New
Hampshire—Mirror Lake (Grafton County).
3. Freshwater biology—New Hampshire—
Mirror Lake (Grafton County). 4. Freshwater
ecology—New Hampshire—Mirror Lake (Grafton
County). 5. Mirror Lake (Grafton County, NH)—
Environmental conditions. I. Winter, Thomas C.
II. Likens, Gene E., 1935–

QH105.N4M55   2009
551.48'2097423—dc22   2009015370

16 15 14 13 12 11 10 09
10 9 8 7 6 5 4 3 2 1

The paper used in this publication meets the
minimum requirements of ANSI/NISO Z39.48-1992
(R 1997)(*Permanence of Paper*).

Cover illustration: Mirror Lake in autumn. Photo by
Donald Buso.

CONTENTS

Contributors   xi

Preface and Acknowledgments   xiii

1. A Limnological Introduction to Mirror Lake   1
   Gene E. Likens
   Limnological History of the Lake   4
   The Lake Today   5
   This Book   13
   Tables   15
   References   19

2. Hydrologic Processes and the Water Budget   23
   Donald O. Rosenberry and Thomas C. Winter
   Hydrogeologic Setting   24
   Hydrologic Processes   27
   Methods of Determining Water Budget Components   31
   Results   44
   Tables   65
   References   67

3. Nutrient Dynamics   69
   *Donald C. Buso, Gene E. Likens, James W. LaBaugh,
   and Darren Bade*
   Research Methods   71
   Research Results   83
   Considerations   128
   Conclusions   152
   Tables   155
   References   201

4. Evaluation of Methods and Uncertainties in the
   Water Budget   205
   Thomas C. Winter and Donald O. Rosenberry
   Water Storage in the Lake   205
   Precipitation   206
   Evaporation   207
   Surface Water   208
   Groundwater   211
   Tables   221
   References   223

5. Evaluation of Methods and Uncertainties in the
   Chemical Budget   225
   James W. LaBaugh, Donald C. Buso, and Gene E. Likens
   Uncertainty in the Water Budgets Used to Determine
      Chemical Budgets   226
   Uncertainties in Chemical Analyses   229
   Uncertainties in Sample Collection   230
   Alternate Approaches to Determining
      Chemical Budgets   236
   The Relation of Uncertainties to Hypotheses   246
   The Relation of Water and Solute Budgets to Lake
      Concentrations   257
   Uncertainty in Perspective   262
   Tables   265
   References   297

6. Mirror Lake: Past, Present, and Future   301
   Gene E. Likens and James W. LaBaugh
   Historical Change   302
   Hydrological and Biogeochemical Fluxes   305
   Management Considerations   308
   Cultural Eutrophication   312
   The Future   316
   References   323

7. Summary and Conclusions   329
   Thomas C. Winter and Gene E. Likens

   Index   335

CONTRIBUTORS

Darren Bade
Kent State University
Kent, Ohio
dbade@kent.edu

Donald C. Buso
Cary Institute of Ecosystem Studies
Millbrook, New York
dbuso@worldpath.net

James W. Labaugh
U.S. Geological Survey
Reston, Virginia
jlabaugh@usgs.gov

Gene E. Likens
Cary Institute of Ecosystem Studies
Millbrook, New York
LikensG@ecostudies.org

Donald O. Rosenberry
U.S. Geological Survey
Lakewood, Colorado
rosenber@usgs.gov

Thomas C. Winter
U.S. Geological Survey
Lakewood, Colorado
tcwinter@usgs.gov

PREFACE

Mirror Lake is a small lake situated at the lower end of the Hubbard Brook Valley in the White Mountains of New Hampshire, USA. Gene Likens, with his colleagues, then of Dartmouth College and Cornell University, and with his students, began limnological studies of the lake in the early 1960s. The work involved studying the physical, chemical, and biological linkages in the lake as well as the linkages of the lake to its watershed and airshed. Much of this work was summarized in a book titled *An Ecosystem Approach to Aquatic Ecology: Mirror Lake and Its Environment*, which was published in 1985. To quote from the preface of that book, "Many of the studies conducted in Mirror Lake were about organisms or communities. Nevertheless, an abiding objective always was to determine the significance that these individuals and individual processes had to the overall structure, metabolism and biogeochemistry of the Mirror Lake Ecosystem." Fundamental to the understanding of ecosystems is knowledge and understanding of the storage and mass balances of water and chemicals of those ecosystems, as well as knowledge of how climate variability and the works of humans affect the mass balances.

In the late 1970s, Mirror Lake was selected by Tom Winter, of the U.S. Geological Survey, for intensive studies of its hydrology. The lake was selected, following a search in the major natural lake regions of the United States, as a field site to establish long-term studies of the interaction of lakes and ground water. The focus of the studies was to be on ground water because it historically had received little attention in lake water budget studies; it commonly was calculated as the residual term. However, from a review of uncertainties in measurement of hydrologic components that commonly are measured, it was clear that the residual also contained all of the errors in the measurements of precipitation, evaporation, and streamflow. As a result, it was decided to instrument Mirror Lake so that all components of the hydrologic system could be measured. In this way, no hydrologic component would be included in the residual, and the uncertainties would have to be dealt with by apportioning them among the individual components. At the same time the hydrology of Mirror Lake was being monitored, chemical fluxes to and from the lake, as well as changes in the storage of chemicals in the lake, were also being measured as part of Likens' long-term studies of the lake. As a result of these activities, we now have a data set that extends for more than 20 years.

This book discusses the water and chemical budgets of Mirror Lake for the 20-year period from 1981 to 2000, a period during which a wide variety of climate conditions affected the lake. Monthly and annual water budgets are presented in chapter 2, and associated chemical budgets in chapter 3. These chapters are followed in chapters 4 and 5 by a lengthy discussion of the uncertainties and difficulties in determining water and chemical budgets. Much was learned about Mirror Lake during the course of these studies that has increased our understanding of the linkages between a lake and its watershed and airshed. Finally, our view and understanding of the past, present, and future of Mirror Lake and the transfer value of this understanding to other lakes is presented in chapter 6.

Studies such as that presented in this book require a long-term commitment not only by the scientists involved but by their funding sources as well. The U.S. Geological Survey has supported these hydrological studies since the inception of the work in 1978. Likens has done chemical, biological, and ecosystem studies on the lake and its airshed and watershed linkages since the early 1960s, with funding from the National Science Foundation, The A.W. Mellon Foundation, Cornell University, and the Institute of Ecosystem Studies.

ACKNOWLEDGMENTS

Long-term financial support for studies of Mirror Lake were provided by the National Science Foundation, including the LTER, LTREB, REU, and RET programs; Dartmouth College; Cornell University; The Andrew W. Mellon Foundation; and the U.S. Geological Survey. Logistical help and support were provided by the U.S. Forest Service and the Hubbard Brook Research Foundation. We acknowledge the efforts of numerous students and colleagues over the years in helping us to unravel the mysteries of air-land-water interactions in the lake and its airshed and watershed.

We also gratefully acknowledge Brent Aulenbach, Ted Stets, David Clow, Greg Noe, Pete VanMetre, Phillip Harte, Terrie Lee, Dennis LeBlanc, Perry Jones, Scott Bailey, and two anonymous reviewers for their reviews of the individual chapters of this book, as well as Sandra Cooper and Keith Lucey for their editorial review of portions of the manuscript. Comments, suggestions, and edits provided by these reviewers and editors allowed us to improve the clarity and presentation of this 20-year episode in the ongoing study of Mirror Lake.

FIGURE 1-1. Mirror Lake in autumn. (Photo by Donald Buso.)

I

A LIMNOLOGICAL INTRODUCTION TO MIRROR LAKE

GENE E. LIKENS

Mirror Lake is a clear-water lake in the Hubbard Brook Valley of the White Mountains of New Hampshire. This beautiful lake has had several names, including Hobart's Pond, McLellan's Pond, Jobert's Pond, Hubbard's Pond, and Tannery Pond, since white settlers colonized the area in the middle 1700s. But on a crisp, clear morning, it is easy to see why this lake has its current name. It reflects its surroundings with the perfection of an expensive mirror (Fig. 1-1). The cultural history of the lake and its drainage basin is interesting and diverse and has included small farms; family, children's, and church camps; a dance hall; a "sugar bush"; a soda bottling operation; saw mills; and a large tannery (Likens 1972; Likens 1985c, pp. 72–83).

Mirror Lake's size and depth are as follows (Table 1-1, Fig. 1-2):

Maximum effective length	610 m
Average depth	5.75 m
Maximum effective width	370 m
Length of shoreline	2.247 km
Area	15.0 ha
Shore development	1.64
Maximum depth	11.0 m
Volume development	1.57
Relative depth	2.5%

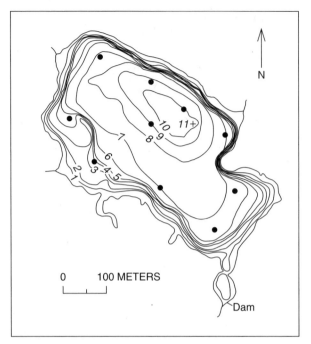

FIGURE 1-2. Bathymetric map of Mirror Lake and locations of thermal surveys. 43°56.5'N, 71°41.5'W. (From Likens et al. 1985, p. 90.)

The single outlet is dammed, but at high lake stages, flow over the dam drains into Hubbard Brook, near the mouth of the Hubbard Brook Valley. Three tributaries flow into the lake from a drainage basin of 103 ha (Winter 1985, pp. 40–53). These streams are simply called NE Tributary, NW Tributary, and W Tributary (Fig. 1-3).

Glacial deposits cover most of the watershed and underlie most of the lake's sediment. A knob of local bedrock (highly variable schist) is exposed along the northeastern shoreline and serves currently as the "swimming rock" for the lake. The maximum relief of the watershed is 268 m.

The climate is humid continental with about 1.4 m of precipitation a year, some 30 percent of which is received as snow. Summer, although

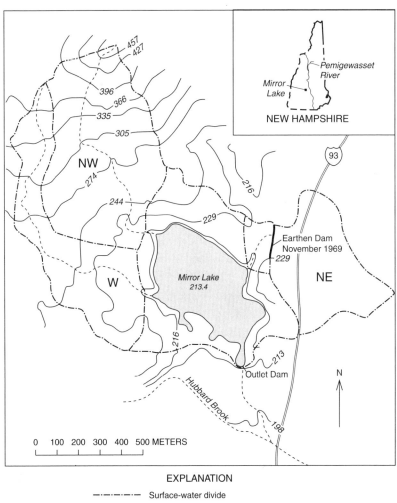

FIGURE 1-3. Outline of Mirror Lake watershed, showing tributaries, subwatershed boundaries, and Interstate 93. 43°56.5'N, 71°41.5'W. The structure on the Northeast Tributary is an earthen dam built prior to the construction of I-93. (From Likens 1985c, p. 81.)

short, is usually hot, and winter, although long, is cold and snowy (Likens 1985a; Likens and Bormann 1995).

In 1969–1971, an interstate highway (I-93) cut through the NE drainage basin of Mirror Lake, diverting the drainage water from about 18 ha of that subdrainage of the lake and thereby reducing the total surface drainage area for the lake to 85 ha (Fig. 1-3). The ecological impact of this interstate highway on the lake and its watershed has been primarily twofold: (1) reduction in water input via the NE Tributary and (2) large input of NaCl as contamination from road salt added to I-93 during the winter for snow and ice removal (Bormann and Likens 1985, pp. 436–444; Rosenberry et al. 1999; Kaushal et al. 2005).

The lake formed from an ice block, a remnant of the retreating glacier. This ice block, buried in the glacial deposits, melted some 14,000 years ago, leaving a depression in the land surface, which then filled with water (Likens and Davis 1975; Davis 1985, pp. 53–65). Currently, Mirror Lake contains some 860,000 m^3 of water (Table 1-1). Water in Mirror Lake is relatively clear and nutrient poor, and is therefore considered oligotrophic (<2 µg/liter chlorophyll-a). Its "high quality" is prized for recreation during the summer, especially for swimming. There is a public beach (Town of Woodstock) on the south shore; boating and fishing are popular, but no gasoline-powered motors are allowed on the lake.

Currently, there are 11 "permanent" residences and a Hamlet of cabins in the Mirror Lake watershed. There are nine housing units in the Hamlet. During the 1980s, about half of the shoreline (northern shore) was purchased for protection against development and currently is owned by the USDA Forest Service. In 2004 the Mirror Lake Hamlet area (southwestern shore) was purchased for shoreline protection, research access, and provision of housing for researchers of the long-term Hubbard Brook Ecosystem Study being done within the Hubbard Brook Valley. Thus, about 70 percent of the shoreline area of Mirror Lake has been protected from further development since 2004.

LIMNOLOGICAL HISTORY OF THE LAKE

Mirror Lake slowly began to fill with sediment some 14,000 years B.P. (before present), when the lake formed. It is estimated from numerous cores of the lake's sediment that the original basin had a maximum depth of about 24 m (Davis and Ford 1985). Thus, the volume of water in the lake today is roughly half of what it was 14,000 years B.P., assuming the water surfaces were similar in ancient and modern times.

A scarcity of charcoal in the sediment profile suggests a relatively low occurrence of fire in the drainage basin during the lake's history. Somewhat more charcoal was accumulated in the sediments 8000 to 7000 years B.P., which correlates with a greater abundance of fire-prone, coniferous vegetation in the landscape at that time (Davis 1985).

From 14,000 to about 200 years B.P., the chemistry and productivity of Mirror Lake changed relatively little (see chapter 6). The lake was cold and relatively unproductive, and the pH was about circumneutral during much of this period. However, the watershed changed appreciably during this time, from a tundra-like landscape, following the retreat of the glacier, to one characterized by coniferous vegetation (e.g., increase in spruce [*Abies* sp.] some 2000 years B.P.), and more recently to one dominated by deciduous vegetation (Davis 1985). Recent human settlement (since about 200 years B.P.) resulted in more rapid changes in chemistry and biology of the lake (see chapter 6). For example, concentrations of cadmium, zinc, lead, iron, and manganese recently have increased in the sediment profile in the lake (Likens and Moeller 1985, pp. 392–410; Sherman 1976). Presumably, the increases in toxic metals in recent sediments are the result of human activities in the lake's watershed and airshed.

THE LAKE TODAY

PHYSICAL LIMNOLOGY

The duration of ice cover on Mirror Lake has become significantly shorter since 1968 (Figs. 1-4 and 1-5; chapter 6). This shortening is correlated with an increase in average air temperature during April of each year (Likens 2000). Because the ice cover on a lake is directly related to its energy budget, changes in duration of the ice cover have been related to regional/global warming (Likens 2000; Magnuson et al. 2000). Generally, the ice cover on Mirror Lake is 40 to 75 cm thick by February, and there may be appreciable snow on it as well (Johnson et al. 1985, pp. 108–127).

Normally, the lake circulates completely ("overturns") throughout the basin twice (dimictic) a year—in spring after the ice cover melts and

FIGURE 1-4. Ice melting on Mirror Lake in April 2002. (Photo by Donald Buso.)

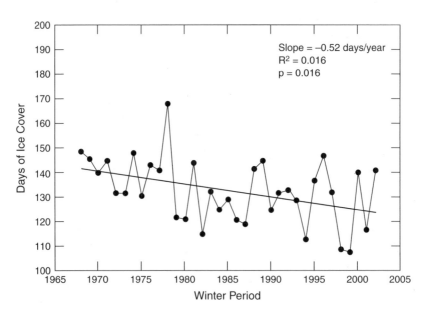

FIGURE 1-5. Long-term record of ice-cover duration on Mirror Lake. (Modified and updated from Likens 2000.)

in fall before the ice cover forms. At these times, the density of water is similar from top to bottom, and the work of the wind acting on the surface of the lake can relatively easily mix the water throughout the entire basin. The lake does not circulate completely during the summer and winter (ice-cover) periods, when the lake is thermally stratified (i.e., has large density differences from top to bottom)(Fig. 1-6; see Johnson et al. 1985, pp. 108–127). The lake is slightly undersaturated with respect to dissolved oxygen and is anoxic below about 8 m during late-summer stratification periods. As the period of ice cover decreases as a result of global warming, the time of summer stratification potentially will be lengthened. As a result, it is likely that anoxia in the hypolimnion will be intensified, with many potential ecological ramifications. These changes in response to global warming are the subject of ongoing investigation.

The theoretical flushing time for water in the lake is about one year, but cold, incoming surface water during the snowmelt period in the spring, which is at a lesser density than the water in the lake, may flow under the ice and out the outlet without mixing into the total volume of the lake, increasing the actual flushing time to 15 months or more (Johnson et al. 1985, pp. 108–127). This topic will be considered in more detail in later chapters.

Water transparency, as measured by a standard (20 cm diameter) Secchi disk, is quite variable throughout the year, but the Secchi depth normally ranges from 5 to 7 m in summer and from 2.5 to 3.5 m in winter. The maximum value recorded was 8.5 m in 1979 (Likens et al. 1985, pp. 89–108). Thus, Mirror Lake is a relatively clear-water lake.

BIOLOGICAL LIMNOLOGY

There are some 850 species of plants, animals, and microorganisms in the lake, but estimates of the number of species in some groups are highly uncertain (Table 1-2). Based on these estimates, benthic invertebrates are the most species rich and represent almost one-half of the total number of species in the lake. It should be noted, however, that appreciable effort has been expended to describe the benthic fauna of Mirror Lake (Walter 1985, pp. 204–228; Strayer 1985a, 1985b), whereas relatively little effort has been spent to describe the species richness of bacteria and fungi. Pelagic algae (phytoplankton) are the next most species-rich group (Table 1-2).

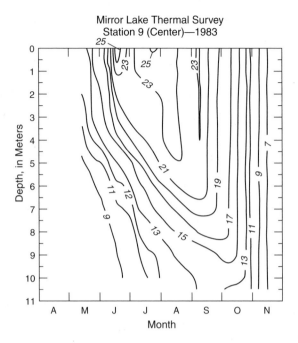
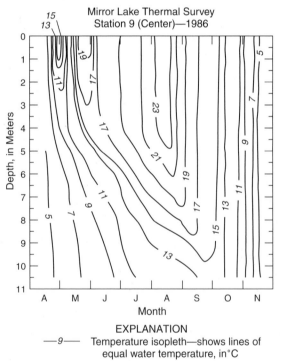

FIGURE 1-6. Depth-time temperature (°C) isopleths for a dry year (1983) and a wet year (1986).

The species diversity in Mirror Lake is not particularly high relative to other lakes. For example, Godfrey (1977) found 217 species of pelagic algae in Cayuga Lake, New York. The relatively low species richness in Mirror Lake may be due to the relatively low nutrient concentrations and slightly acidic conditions in the water, the relatively long period of ice cover, and the relative absence of macrophytes, all of which reduce habitat diversity and all of which are characteristic of oligotrophic lakes in the North Temperate Zone.

There are very few aquatic macrophytes in Mirror Lake in terms of either species or abundance (Moeller 1985, pp. 177–192). Small patches of water lilies (*Nuphar* and *Nymphaea* spp.), burr weed (*Sparganium* sp.), pipewort (*Eriocaulon* sp.), and water lobelia (*Lobelia* sp.) can be seen at the surface of the lake in summer. Below the surface, patches of bladderwort (*Utricularia* sp.), pondweeds (*Potamogeton* sp.), *Isoetes* sp., stoneworts (*Nitella* sp.), and a few other submerged plants occur. At times during the summer, bladderwort may become quite abundant in some areas of the lake. The scarcity of macrophytes in Mirror Lake may be due to the low nutrient content of the lake and the large expanses of sand, cobbles, and boulders in shallow areas (Fig. 1-7), but the definitive answer has not been determined (Moeller 1985).

Until recently, Mirror Lake contained five species of fish: smallmouth bass (*Micropterus dolomieui*), yellow perch (*Perca flavescens*), chain pickerel (*Esox niger*), white sucker (*Catostomus commersoni*), and brown bullhead (*Ictalurus nebulosus*). In addition to this low diversity, none of these species occur in large numbers (Mazsa 1973; Helfman 1985). In 1995, the New Hampshire Fish and Game Department stocked rainbow trout (*Oncorhynchus mykiss*), brown trout (*Salmo trutta*), and brook trout (*Salvelinus fontinalis*) in the lake and has continued to stock rainbow trout since. As a result, the recreational fishery in the lake has increased greatly in popularity, especially in winter (Fig. 1-8).

A small population of the red-spotted newt (*Notophthalmus v. viridescens*) inhabits Mirror Lake and often can be seen in tributary areas as well (Burton 1985). These newts normally experience low predation because of toxins in their skin. Other amphibians (e.g., the green frog [*Rana clamitans*] and the American toad [*Bufo americanus*]) inhabit shoreline areas. Also, painted turtles (*Chryscarys picta*) may be seen basking, and common snappers (*Chelydra serpentina*) are found, although rarely. The common loon (*Gavia immer*), typically immature and solitary, is present, as are mallard ducks.

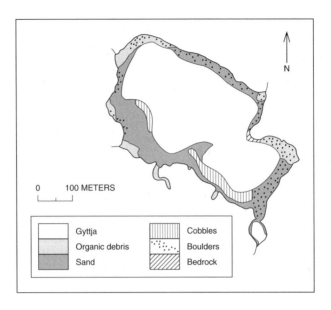

FIGURE 1-7. Distribution of lakebed sediments in Mirror Lake.

Overall, Mirror Lake is relatively unproductive. About 0.34 percent of incoming solar radiation is fixed annually by photosynthetic plants in Mirror Lake. Phytoplankton productivity accounts for about 90 percent of total plant productivity on an annual basis in Mirror Lake, as rooted macrophytes are relatively rare (Table 1-3). Some 70 percent of all inputs of organic carbon to the lake are respired by consumers (Likens 1985b).

CHEMICAL LIMNOLOGY

The chemistry of Mirror Lake, as well as its three tributaries, is currently dominated by calcium and sulfate (Table 1-4), although concentrations of sodium and chloride have been increasing in the NE Tributary and in the lake because of road salt contamination from I-93 (see chapters 3 and 6). Concentrations of nutrients (phosphorus and nitrogen), usually limiting to biological productivity in freshwater ecosystems, are very low in Mirror Lake (Table 1-4) and co-limit biological productivity (Gerhart and Likens 1975; Bade et al. 2008). Currently, pH averages around 6.4, acid-neutralizing capacity (ANC) is 81 mg/liter, average specific conductivity is about 31 µS/cm, and average dissolved organic carbon (DOC) concentration is approximately 1.8 mg/liter (Table 1-4). A more detailed

FIGURE 1-8. So-called "ice houses" used for fishing on Mirror Lake in January 2003. (Photo by Gene E. Likens.)

analysis and a discussion of long-term trends in the lake's chemistry are presented in chapters 3 and 6.

A budgetary or mass balance approach has been used successfully for many decades as part of the Hubbard Brook Ecosystem Study to analyze biogeochemical and hydrological fluxes and change (see Bormann and Likens 1967; Likens and Bormann 1995; Likens 1992). In the most simple form, the mass balance for an ecosystem can be described as Inputs = Outputs + Δ Storage. For Mirror Lake, this equation can be expanded to

$$P + GWI + SWI + GASIN$$
$$= SWO + GWO + GASOUT + \Delta \text{ Storage} \quad (1\text{-}1)$$

where

P is direct bulk precipitation on the lake's surface,
GWI is groundwater seepage to the lake,
SWI is tributary stream water,

SWO is flow over the dam at the lake outlet,
GWO is seepage of lake water to ground water,
$GASIN$ and $GASOUT$ are water vapor and elements in gaseous form at normal biological temperatures (e.g., N_2),
Storage includes long-term biological uptake and sedimentation.

All terms in Equation 1-1 are subject to uncertainty, which is considered in chapters 4 and 5. Equation 1-1 can be considered as a net ecosystem budget or balance (see Likens et al. 2002).

This approach has been used to estimate ecosystem parameters that are difficult to measure, such as weathering and evapotranspiration in watershed ecosystems of the Hubbard Brook Experimental Forest (Bormann and Likens 1967). It will be the basis for much of the analysis done in chapters 2, 3, 4, and 5.

The average standing stock of carbon, nitrogen, and phosphorus in Mirror Lake is given in Table 1-5. Most of the carbon, nitrogen, and phosphorus is located in the top 10 cm of sediment, followed by the dissolved

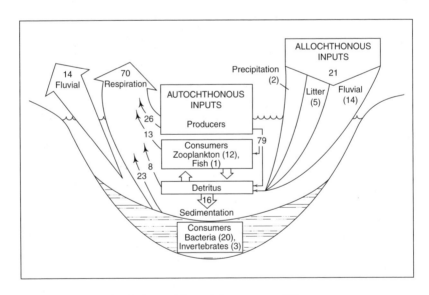

FIGURE 1-9. Food web for the Mirror Lake ecosystem. Values represent percent of organic carbon flux through Mirror Lake. Sedimentation represents net sedimentation. (From Likens 1985b, pp. 337–344.)

component. The completeness of the total phosphorus budget for Mirror Lake is unusual (Cole et al. 1990). Movement of organic carbon through the food web of Mirror Lake is diagrammed in Fig. 1-9.

THIS BOOK

Mirror Lake has been the focus of comprehensive and continuous scientific study since 1965 and as such is one of the most studied lakes in the world. These studies led in 1985 to a comprehensive book, *An Ecosystem Approach to Aquatic Ecology: Mirror Lake and Its Environment* (Likens 1985a). Much of this book was the product of my own research, done with the help of my graduate students, postdocs, technicians, and colleagues. The geological and cultural history of Mirror Lake, as well as its ecology, paleoecology, biogeochemistry, and limnology (study of inland waters), are discussed in detail, but several of the hydrologic components (e.g., evaporation and deep seepage of ground water) were determined by approximation or difference. Here, the magnitude of these terms is determined directly.

Limnologically, Mirror Lake is very similar to many lakes in northern New England and the Precambrian Shield of Canada with respect to biogeochemistry, hydrology, and limnology (Likens 1985b). The previous book about Mirror Lake (Likens 1985a) attempted to detail many of the processes, functions, and species (biotic and abiotic) that have been identified and studied in the lake, and therefore provided a framework for understanding the lake as the sum of its working parts: an aquatic ecosystem. That effort was, in effect, a collection of linked puzzle pieces, but quantitatively, it lacked some of the edges for the completed scene. This approach seems counterintuitive, given that the easy part of any puzzle is supposed to be the edges. Here is the paradox: Ecosystems are difficult to study because defining the pieces can require much more information than can be gathered in just a few field seasons, even when focusing on central processes or critical species. Guessing at how the edges fit together leaves many important questions unsolvable. Thus, this second book is all about quantifying two critical pieces of the Mirror Lake ecosystem puzzle, the hydrology and chemistry of the lake. These components are intricately and inescapably linked to allow the ecosystem to function. Just why these pieces are important, and how they can be measured quantitatively, is the focus of this book. The quantitative evaluation of the hydrologic and chemical pieces (quantity and flux) of the lake allow more of the ecosystem puzzle to be assembled, and constrain as the matrix all other ecosystem "pieces" and functions, large or small.

This current book provides a detailed analysis of the hydrology and chemistry of this small lake and focuses on the long-term linkages among air, land, and water during 1981 to 2000, with a uniquely detailed and complete data set. Data from these long-term studies provide important insights into the function and change in this aquatic ecosystem, as well as about the lake's linkages with regional and global systems.

TABLE 1-1. *Morphometric and volumetric characteristics of Mirror Lake*

Depth (m)	Area		Stratum (m)	Volume	
	$m^2 \times 10^4$	% of total		$m^3 \times 10^3$	% of total
0	15.0	100.0	0–1	142.9	16.6
1	13.6	90.5	1–2	130.0	15.1
2	12.4	82.9	2–3	119.5	13.9
3	11.5	76.5	3–4	110.0	12.8
4	10.5	70.1	4–5	101.8	11.8
5	9.86	65.7	5–6	94.1	10.9
6	8.96	59.7	6–7	78.5	9.1
7	6.79	45.2	7–8	48.9	5.7
8	3.21	21.4	8–9	23.6	2.7
9	1.61	10.7	9–10	10.7	1.2
10	0.609	4.06	10–11	2.0	0.2
11	0	0			
			Total	862.0	100.0

SOURCE: Modified from Winter 1985.

TABLE 1-2. *Estimated number of species in Mirror Lake*

Type	Number of species[a]
Pelagic algae (phytoplankton)	138
Benthic algae	>50?
Macrophytes	37
Pelagic bacteria	>50??
Pelagic fungi	>10???
Benthic bacteria	>100??
Benthic fungi	>10???
Pelagic zooplankton (includes Protozoa)	>50?
Benthic invertebrates	>400?
Fish	6
Reptiles and amphibians	4–7
Birds	4–5
Mammals	2–5
Total	>850

SOURCE: From Likens 1992.

[a] ? indicates relative uncertainty.

TABLE 1-3. *Inputs of organic carbon for Mirror Lake*

Source	Carbon ($mg\ m^{-2}\ yr^{-1}$)
Autochthonous (gross)	
Phytoplankton (POC and DOC)	56,500[a]
Epilithic algae	2500[b]
Epipelic and epiphytic algae	>1000[b]
Macrophytes	2500[b]
Dark CO_2 fixation	2100[b]
	Total 64,600
Allochthonous	
Precipitation	1400[c]
Shoreline litter	4300[c]
Fluvial[d]	
DOC	10,500[e]
FPOC (0.45 μm–1 mm)	300[e]
FPOC (>1 mm)	50[e]
CPOC	800[e]
	Total 17,350
	Combined inputs 81,950

SOURCE: From Likens 1985b. Carbon data revised from Jordan and Likens 1975.

[a] Daytime ^{14}C fixation = 47,000 (POC = 38,300; DOC = 8700). Gross POC fixation = 47,800. Net POC fixation = 28,700 (60% × 47,800). Day and night respiration = 19,100 (40% × 47,800). Net POC = 0.75 × [^{14}C]POC. The 0.75 is a correction for nighttime respiration based on net = 60% gross + R = 10% P_{max} (Steemann Nielsen 1958).

[b] Right order of magnitude.

[c] +20%.

[d] DOC = dissolved organic carbon. FPOC = fine particulate organic carbon. CPOC = course particulate organic carbon.

[e] +50%.

TABLE 1-4. *Average chemical data for Mirror Lake, 1981 to 2000 (20 years)*

Chemical	Lake[a] average	Lake[a] range	W tributary average	W tributary range
Calcium	2.39	3.47–0.73	2.62	6.57–1.27
Sodium	2.11	3.53–0.66	2.36	5.87–0.95
Magnesium	0.51	0.65–0.23	0.57	1.37–0.21
Potassium	0.46	0.64–0.18	0.54	1.66–0.18
Ammonium	0.02	0.22–0.01	0.02	0.13–0.01
H^+	0.4	1.9–0.1	1.1	7.9–0.5
Aluminum	0.01	0.010–0.005	0.01	0.10–0.01
Sulfate	4.52	16.2–1.6	5.08	11.7–0.7
Chloride	2.86	5.19–1.02	2.65	7.41–0.70
Nitrate	0.04	0.75–0.01	0.16	2.28–0.01
Phosphate	0.009	0.080–0.001	0.009	0.092–0.001
Dissolved silicate	2.07	3.7–0.1	9.2	15.2–3.7
Dissolved organic C	1.8	3.4–1.5	3.1	16.2–1.0
pH	6.38	7.00–5.72	6.13	7.15–5.10
Sp. cond. ($\mu S\ cm^{-1}$)	31	45–22	31	55–16
ANC	81	270–7	90	341–(–13)
Dissolved inorganic C	163	804–55	101	393–34
Dissolved O (% sat)	84	145–39	N/A	—

NOTE: Units in mg L^{-1} unless noted otherwise.

[a] Lake water for depths of 2–8 m, not from surface or hypolimnion, to avoid dilution from ice melt during spring and anoxia during summer stratified periods. To convert dissolved silicate to silica, multiply by 0.47.

TABLE 1-5. *Average standing stock and ratios (weight basis) of carbon, nitrogen, and phosphorus in the Mirror Lake ecosystem*

Component	kcal × 10^{3a}	C (kg)[b]	N (kg)[b]	P (kg)[b]	C : N : P (by wt)
Dissolved[c]	33,600	3360[d]	124	2.5	1344 : 50 : 1
Seston	2420	247[e]	32.3	2.48	100 : 13 : 1
Bacterioplankton	120	12[f]	2.1[g]	0.3[g]	40 : 7 : 1
Phytoplankton	411	56[d]	9.8[g]	1.4[g]	40 : 7 : 1
Epilithic algae	570	57[f]	3.0	0.075	760 : 40 : 1
Macrophytes	2020	202[h]	14	0.8	252 : 17 : 1
Zooplankton	313	30[d]	4.9	0.84	36 : 5.8 : 1
Fish	408	36[f]	8.2	2.7	13.1 : 3.0 : 1
Salamanders	10	0.77[h]	0.13	0.055	14.1 : 2.4 : 1
Benthic macroinvertebrates	1050	105[h]	25.5	3.0	35 : 8.5 : 1
Benthic bacteria	6000	600[f]	105[g]	15[g]	40 : 7 : 1
Sediment (top 10 cm)	733,000	150,000[d]	10,570	1280	117 : 8.3 : 1

SOURCE: Modified from Likens 1985b.

[a] Computed by converting C to dry weight (C ÷ 0.45) and multiplying by mean caloric values from Cummins and Wuycheck (1971).

[b] Divide by 15 × 10^{-4} to convert to kcal/m^2 or by 0.15 to convert to mg/m^2.

[c] Organic plus inorganic: DIC = 1290 kg, DOC = 2070 kg, NH_4-N = 21.6 kg, NO_3-N = 16.4 kg, organic N = 86 kg, Total PO_4-P = 1.7 kg.

[d] +20%.

[e] Excluding zooplankton.

[f] Right order of magnitude.

[g] Assuming a weight ratio of 40 : 7 : 1 for C : N : P.

[h] +50%.

REFERENCES

Bade, D.L., K. Bouchard, and G.E. Likens. 2008. Algal co-limitation by nitrogen and phosphorus persists after 30 years in Mirror Lake (New Hampshire, USA). *Verh. Internation. Verein. Limnol.* 30 (in press).

Bormann, F.H., and G.E. Likens. 1967. Nutrient cycling. *Science* 155(3761):424–429.

Bormann, F.H., and G.E. Likens. 1985. Air and watershed management and the aquatic ecosystem. In G.E. Likens (ed.). *An Ecosystem Approach to Aquatic Ecology: Mirror Lake and Its Environment* (pp. 436–444). New York: Springer-Verlag.

Burton, T.M. 1985. Production and limiting factors: Vertebrates. In G.E. Likens (ed.). *An Ecosystem Approach to Aquatic Ecology: Mirror Lake and Its Environment* (pp. 288–291). New York: Springer-Verlag.

Cole, J.J., N.F. Caraco, and G.E. Likens. 1990. Short-range atmospheric transport: A significant source of phosphorus to an oligotrophic lake. *Limnol. Oceanogr.* 35(6):1230–1237.

Cummins, K.W., and J.C. Wuycheck. 1971. Caloric equivalents for investigations in ecological energetics. *Mitt. Int. Ver. Theore. Angew. Limnol.* 18. 158 pp.

Davis, M.B. 1985. History of the vegetation on the Mirror Lake watershed. In G.E. Likens (ed.). *An Ecosystem Approach to Aquatic Ecology: Mirror Lake and Its Environment* (pp. 53–65). New York: Springer-Verlag.

Davis, M.B., and J. Ford. 1985. Late-glacial and Holocene sedimentation. In G.E. Likens (ed.). *An Ecosystem Approach to Aquatic Ecology: Mirror Lake and Its Environment* (pp. 345–355). New York: Springer-Verlag.

Gerhart, D.Z., and G.E. Likens. 1975. Enrichment experiments for determining nutrient limitation: Four methods compared. *Limnol. Oceanogr.* 20(4):649–653.

Godfrey, P.J. 1977. Spatial and temporal variation of the phytoplankton in Cayuga Lake. Ph.D. dissertation, Cornell University. 512 pp.

Helfman, G. 1985. Fishes. In G.E. Likens (ed.). *An Ecosystem Approach to Aquatic Ecology: Mirror Lake and Its Environment* (pp. 236–245). New York: Springer-Verlag.

Johnson, N.M., G.E. Likens, and J.S. Eaton. 1985. Stability, circulation and energy flux in Mirror Lake. In G.E. Likens (ed.). *An Ecosystem Approach to Aquatic Ecology: Mirror Lake and Its Environment* (pp. 108–127). New York: Springer-Verlag.

Jordan, M.J., and G.E. Likens. 1975. An organic carbon budget for an oligotrophic lake in New Hampshire, U.S.A. *Verh. Internat. Verein. Limnol.* 19(2):994–1003.

Kaushal, S.S., P.M. Groffman, G.E. Likens, K.T. Belt, W.P. Stack, V.R. Kelly, L.E. Band, and G.T. Fisher. 2005. Increased salinization of fresh water in the northeastern United States. *Proc. National Academy of Sciences* 102(38):13517–13520.

Likens, G.E. 1972. Mirror Lake: Its past, present and future? *Appalachia* 39(2):23–41.

Likens, G.E. (ed.). 1985a. *An Ecosystem Approach to Aquatic Ecology: Mirror Lake and Its Environment*. New York: Springer-Verlag. 516 pp.

Likens, G.E. 1985b. The lake ecosystem. In G.E. Likens (ed.). *An Ecosystem Approach to Aquatic Ecology: Mirror Lake and Its Environment* (pp. 337–344). New York: Springer-Verlag.

Likens, G.E. 1985c. Mirror Lake: Cultural history. In G.E. Likens (ed.). *An Ecosystem Approach to Aquatic Ecology: Mirror Lake and Its Environment* (pp. 72–83). New York: Springer-Verlag.

Likens, G.E. 1992. *The Ecosystem Approach: Its Use and Abuse*. Excellence in Ecology, Book 3. Oldendorf-Luhe, Germany: The Ecology Institute. 166 pp.

Likens, G.E. 2000. A long-term record of ice cover for Mirror Lake, New Hampshire: Effects of global warming? *Verh. Internat. Verein. Limnol.* 27(5):2765–2769.

Likens, G.E., and F.H. Bormann. 1995. *Biogeochemistry of a Forested Ecosystem*, . Second Edition. New York: Springer-Verlag. 159 pp.

Likens, G.E., and M.B. Davis. 1975. Post-glacial history of Mirror Lake and its watershed in New Hampshire, USA: An initial report. *Verh. Internat. Verein. Limnol.* 19(2):982–993.

Likens, G.E., C.T. Driscoll, D.C. Buso, M.J. Mitchell, G.M. Lovett, S.W. Bailey, T.G. Siccama, W.A. Reiners, and C. Alewell. 2002. The biogeochemistry of sulfur at Hubbard Brook. *Biogeochemistry* 60(3):235–316.

Likens, G.E., J.S. Eaton, and N.M. Johnson. 1985. Physical and chemical environment. In G.E. Likens (ed.). *An Ecosystem Approach to Aquatic Ecology: Mirror Lake and Its Environment* (pp. 89–108). New York: Springer-Verlag.

Likens, G.E., and R.E. Moeller. 1985. Chemistry. In G.E. Likens (ed.). *An Ecosystem Approach to Aquatic Ecology: Mirror Lake and Its Environment* (pp. 392–410). New York: Springer-Verlag.

Magnuson, J.J., D.M. Robertson, B.J. Benson, R.H. Wynne, D. Livingstone, T. Arai, R.A. Assel, R.G. Barry, V. Card, E. Kuusisto, N.G. Granin, T.D. Prowse, K.M. Stewart, and V.S. Vuglinski. 2000. Historical trends in lake and river ice cover in the Northern Hemisphere. *Science* 289:1743–1746.

Mazsa, D. 1973. The ecology of fish populations in Mirror Lake, New Hampshire. M.S. thesis, Cornell University. 172 pp.

Moeller, R.E. 1985. Species composition, distribution, population, biomass and behavior: Macrophytes. In G.E. Likens (ed.). *An Ecosystem Approach to Aquatic Ecology: Mirror Lake and Its Environment* (pp. 177–192). New York: Springer-Verlag.

Rosenberry, D.O., P.A. Bukaveckas, D.C. Buso, G.E. Likens, A.M. Shapiro, and T.C. Winter. 1999. Movement of road salt to a small New Hampshire lake. *Water, Air, and Soil Pollut.* 109:179–206.

Sherman, J.W. 1976. Post-Pleistocene diatom assemblages in New England lake sediments. Ph.D. dissertation, University of Delaware. 312 pp.

Steeman Nielsen, E. 1958. Experimental methods for measuring organic production in the sea. *Rapp. P.-V. Reun. Cons. Int. Explor. Mer* 144:38–46.

Strayer, D.L. 1985a. Benthic microinvertebrates. In G.E. Likens (ed.). *An Ecosystem Approach to Aquatic Ecology: Mirror Lake and Its Environment* (pp. 228–234). New York: Springer-Verlag.

Strayer, D.L. 1985b. The benthic micrometazoans of Mirror Lake, New Hampshire. *Arch. Hydrobiol./Suppl.* 72(3):287–426.

Walter, R.A. 1985. Species composition, distribution, population, biomass and behavior: Benthic macroinvertebrates. In G.E. Likens (ed.). *An Ecosystem Approach to Aquatic Ecology: Mirror Lake and Its Environment* (pp. 204–228). New York: Springer-Verlag.

Winter, T.C. 1985. Physiographic setting and geologic origin of Mirror Lake. In G.E. Likens (ed.). *An Ecosystem Approach to Aquatic Ecology: Mirror Lake and Its Environment* (pp. 40–53). New York: Springer-Verlag.

2

HYDROLOGIC PROCESSES AND THE WATER BUDGET

DONALD O. ROSENBERRY AND
THOMAS C. WINTER

Mirror Lake was selected for intensive study of its hydrology following a search in the major, natural lake regions of the United States to establish long-term studies of the interaction of lakes and ground water. Mirror Lake represents a typical mountain lake in a humid continental climate. The original focus of the long-term lake studies was on ground water because it historically had received little attention in water budget studies of lakes; it commonly is calculated as the residual term. A review of uncertainties in measurement of hydrologic components that commonly are measured (Winter 1981) made it clear that the residual also contains all of the errors in the measurements of precipitation, evaporation, and streamflow. As a result, it was decided to instrument and measure all of the hydrologic components of Mirror Lake, including ground water. In this way, no hydrologic component would be included in the residual, and the uncertainties would have to be dealt with by apportioning them among the individual components.

To provide some background to the detailed discussion of Mirror Lake's water budget, this chapter begins with a description of the hydrogeologic setting of the lake. The hydrogeologic setting section extends description of the physical setting of the lake to the subsurface and complements the description of the physiographic setting of the lake that is presented in chapter 1. A brief overview of the hydrologic processes associated with Mirror Lake is provided in the next section of the chapter, followed by results from the hydrologic studies, which are based on monthly and annual water budgets for the calendar years 1981 through 2000.

HYDROGEOLOGIC SETTING

The Hubbard Brook Valley was carved out of crystalline bedrock by glacial advances that occurred during the last 100,000 years of the Pleistocene epoch (Winter 1985). Throughout most of the valley, the bedrock is overlain by thin (from zero to as much as several meters) glacial deposits that consist primarily of till, an unsorted mixture of clay through boulder-sized rocks (Fig. 2-1). At the lower end of the valley, in the vicinity of Mirror Lake, the glacial deposits are much thicker in places, as much as 55 m at the ridge north of the lake (Fig. 2-2). Part of the great thickness of glacial deposits at this location is due to the fact that they fill a buried bedrock valley that trends north from its head beneath the south shore of Mirror Lake (Fig. 2-3). Another buried bedrock valley extends south from the saddle beneath the south shore

FIGURE 2-1. Glacial till along the drainage divide north of Mirror Lake. (Photo by Thomas Winter.)

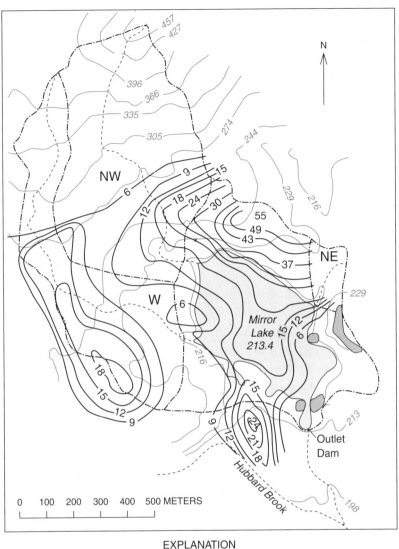

EXPLANATION

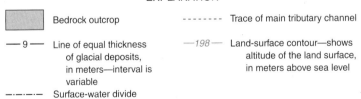

FIGURE 2-2. Thickness of glacial deposits in the Mirror Lake area.

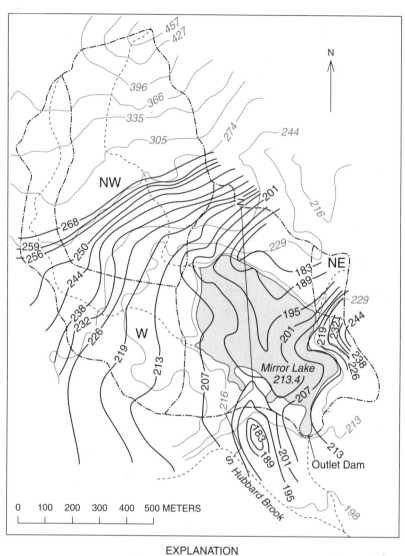

EXPLANATION

—207— Bedrock contour—shows altitude of bedrock surface, in meters above sea level. Interval is variable

N——S Line of section shown in Figure 2-4

—·—·— Surface-water divide
-------- Trace of main tributary channel
—198— Land-surface contour—shows altitude of the land surface, in meters above sea level

FIGURE 2-3. Bedrock topography in the Mirror Lake area.

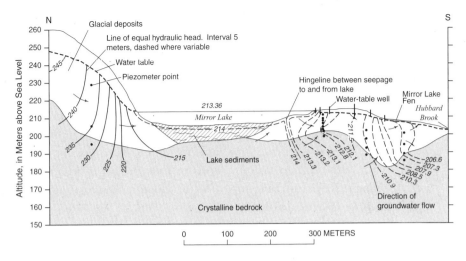

FIGURE 2-4. Hydrogeologic cross-section through the Mirror Lake area. Transect shown in Fig. 2-3.

of the lake. This southern bedrock valley is filled with sand and gravel that was deposited by Hubbard Brook when the stream was adjusting to a higher base level of the Pemigewasset River at the end of the last ice age. During this time, an ice block must have been present at the future site of Mirror Lake, and it was the melting of this ice block that formed the depression that holds Mirror Lake (see chapter 1; Winter 1985).

Although the glacial deposits are quite thick at places near Mirror Lake, bedrock outcrops exist at several locations along the east and southeast shore of the lake (Fig. 2-2). In addition, based on geophysical mapping and coring of lake sediments, it is likely that some lake sediments near the south shoreline are in direct contact with the bedrock (Fig. 2-4). As a result of its hydrogeologic setting, it is evident that Mirror Lake interacts primarily with ground water flowing within and streams flowing upon glacial deposits, and that very little of its water is derived from bedrock sources (Rosenberry and Winter 1993; Tiedeman et al. 1997).

HYDROLOGIC PROCESSES

Observers of lakes commonly perceive them as static features because they look much the same each time they are visited. However, a lake is actually very dynamic, receiving new water from precipitation, stream inflow, and groundwater inflow, and losing water to evaporation, stream outflow, and outflow to ground water. These gains and losses of water affect the amount of water that is present (or stored) in the lake at any one time.

Precipitation is the ultimate source of all water to Mirror Lake and its watershed. Some precipitation falls directly on the lake surface, which adds directly to the lake's water budget. However, a much greater amount falls on the lake's watershed, simply because the watershed covers a much larger area than the lake. In the case of Mirror Lake, the watershed area, not including the lake surface, is 103 ha, 6.9 times larger than the surface of Mirror Lake (Winter 1985). Precipitation that falls on the watershed follows several flow paths that may or may not add to the lake's water budget. A large amount of precipitation that falls on the watershed is returned to the atmosphere by evaporation from wet surfaces and by transpiration from plants before it reaches the lake. Although this terrestrial evapotranspiration was not measured as part of the study of Mirror Lake, it represents about 38 percent of the water that originally falls on the watershed (Likens and Bormann 1995). The remainder of the precipitation either runs off to streams by way of overland flow or shallow subsurface flow, or it infiltrates through the unsaturated zone to recharge ground water. A small amount also runs off directly to the lake, but that volume is assumed to be so small as to be irrelevant to the lake water budget and is ignored (Likens and Bormann 1995).

The three tributary streams that drain the northwest (NW), west (W), and northeast (NE) watersheds of Mirror Lake contribute the largest amount of water to the lake's water budget. Because of the steep slopes, the presence of macropores in the shallow subsurface associated with extensive forest cover and tree roots (Likens and Bormann 1995), and the low transient storage capacity of the glacial till, the tributaries respond rapidly to precipitation, resulting in discharge hydrographs that have sharp peaks and steep recessions (Fig. 2-5). Although the glacial till has low transient storage capacity, it does provide some storage, and it is the release of ground water from the glacial till that sustains streamflow beyond several days following precipitation events.

Ground water also provides direct input to the lake wherever the elevation of the groundwater surface (i.e., the water table) is higher than the lake surface, which is the case for most of the lake except on the south

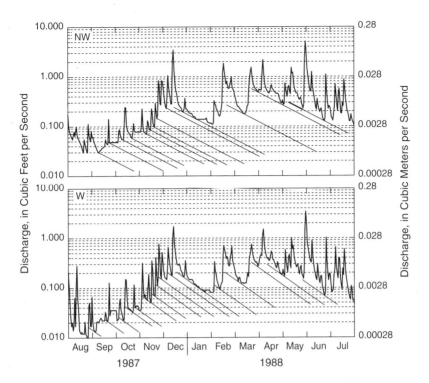

FIGURE 2-5. Hydrographs of discharge of the Northwest Tributary and the West Tributary.

side (Fig. 2-6). Most of the groundwater discharge occurs in the shallow, near-shore margins of the lake. Even though glacial till underlies most of Mirror Lake, ground water contributes the smallest amount of water to the lake's water budget because the glacial till has low permeability. Nevertheless, dissolved chemicals brought into the lake by way of ground water are an important component of the lake's nutrient budget (chapter 3).

The seepage of lake water to ground water is the largest loss of water from Mirror Lake on an annual basis. This loss makes Mirror Lake unusual because surface water outflow is the largest loss term for most lakes that have a surface water outlet. Unlike groundwater inflow, which is slowly released from the poorly permeable glacial till, seepage outflow takes place through highly permeable sand and gravel that fill the buried bedrock valley south of the lake, where the adjacent water table is lower than the lake water surface (Fig. 2-6). Variability in seepage losses

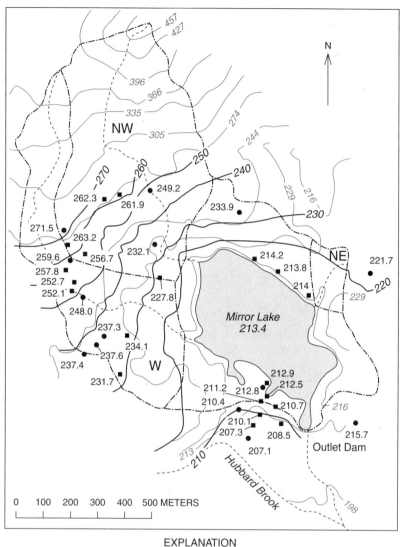

FIGURE 2-6. Configuration of the long-term average altitude of the water table in the Mirror Lake area.

is directly related to changes in lake level and the water table elevation downgradient from the lake. Some lake water also is lost via seepage through fractures in the bedrock, but based on extensive investigations of flow through fractured bedrock within the Mirror Lake watershed (Shapiro et al. 2007), we assume that this volume is very small relative to flow through the unconsolidated sand and gravel.

Surface outflow over the dam is the second largest loss from Mirror Lake on an annual basis. However, during some seasons, primarily in spring and late fall, surface outflow is much larger than seepage outflow. When the lake surface is below the top of the dam, a condition that can last for months at a time, surface water outflow is minimal, occurring as leakage between the boards that compose the upper portion of the dam.

Evaporation from Mirror Lake's surface is the smallest loss of water from the lake. Driven primarily by solar and atmospheric radiation, evaporation generally follows a seasonal cycle from minimal values during fall, winter, and spring to maximum values during late July and early August (Winter et al. 2003; Rosenberry et al. 2007). Transpirational loss from emergent macrophytes is negligible in Mirror Lake because of the sparseness of these plants (see chapter 1).

METHODS OF DETERMINING WATER BUDGET COMPONENTS

This section presents information on the field and analytical methods that were used to determine the water budget of Mirror Lake. Water storage in the lake is presented first because the lake is the "accounting unit" or "reference volume" of the budget. It provides an integrated response to all of the hydrologic fluxes that exchange water with the lake. The method for determining water storage in the lake is followed by the three components of the hydrologic system interacting with the lake: atmospheric water, surface water, and ground water. Locations of instruments are shown in Fig. 2-7.

WATER STORAGE IN THE LAKE

Continuous records of lake stage were collected using an analog strip-chart recorder that was attached to a float and counterweight system mounted inside a stilling well near the lake outlet (Fig. 2-8). A stage-volume relation was developed to provide a continuous record of changes in the water

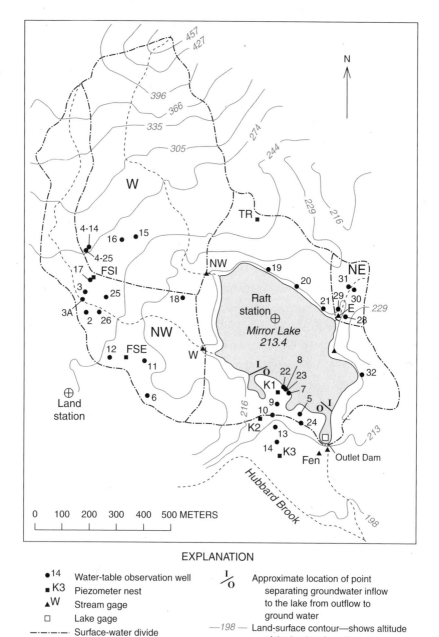

EXPLANATION

- •14   Water-table observation well
- ■K3 Piezometer nest
- ▲W   Stream gage
- ☐   Lake gage
- —··—··— Surface-water divide
- ------ Trace of main tributary channel

$\frac{I}{O}$ Approximate location of point separating groundwater inflow to the lake from outflow to ground water

—198— Land-surface contour—shows altitude of the land surface, in meters above sea level

FIGURE 2-7. Locations of instruments to measure hydrologic variables in the Mirror Lake area.

FIGURE 2-8. Mirror Lake dam and lake stage gauge. The recorder is mounted in the box on top of the corrugated-steel stilling well. (Photo by Thomas Winter.)

volume of the lake. However, because lake stage fluctuated over a narrow range of less than half a meter during the study, and because much of the shoreline of Mirror Lake is nearly vertical at this normal stage, a constant lake surface area of 150,000 m^2 was used when determining change in lake volume associated with change in lake stage. Daily change in lake volume was determined by calculating the difference between successive daily average lake stages and multiplying that difference by the lake surface area. Monthly and annual changes in lake volume were determined by summing daily changes in lake volume during each month or year.

ATMOSPHERIC WATER

Precipitation

Precipitation was measured by a recording weighing-bucket gauge and a standard manually read volumetric rain gauge (Federer et al. 1990; Buso et al. 2000)(Fig. 2-9) at two locations near the lake. One pair of gauges was located about 0.4 km west of Mirror Lake at the U.S. Forest

FIGURE 2-9. Precipitation gauges in forest opening near the U.S. Forest Service Robert S. Pierce Ecosystem Laboratory. (Photo by Thomas Winter.)

Service Robert S. Pierce Ecosystem Laboratory site, referred to here as the land station (Figs. 2-7 and 2-9), and the other was located about 0.5 km southeast of the lake at the Pleasant View Farm site (east of the area shown in Fig. 2-7). Precipitation input to the lake for each event was calculated by multiplying the average of values measured at these two sites by the surface area of the lake.

Evaporation

Evaporation from Mirror Lake was determined by the Priestley-Taylor method. This method was selected after comparing 14 methods with the Bowen-Ratio Energy Budget (BREB) method (Rosenberry et al. 2007; see chapter 4). The BREB method was used at Mirror lake from 1982 to 1987 (Winter et al. 2003) because it was considered to be the most accurate method for determining evaporation, and we intended it to be the standard against which other methods requiring less cost and manpower could be compared. Therefore, the following describes the instruments used for the BREB method as well as for several other alternate methods that were considered.

Instruments for determining evaporation from Mirror Lake were located at land and raft stations (Fig. 2-7). Over the 20 years of this study, some instruments were improved, and new ones were invented. In an attempt to determine the best possible values for any given variable, we tried to keep pace with new technology. Whenever an improved or new instrument was installed at the site, the old and new instruments were operated side by side for at least one year, either at Mirror Lake or at another data collection site that was running concurrently, to assure the new instrument was at least as good as the old and to determine any bias associated with instrument upgrades.

At the raft station near the middle of the lake (Fig. 2-10), during the early part of the study, the primary instruments were (1) anemometers positioned at 1, 2, and 3 m above the water surface to measure wind speed; (2) a thermistor positioned beneath the raft, submerged within 1 cm of the lake surface, to measure water surface temperature; and (3) a thermistor psychrometer positioned at 2 m above the water surface to measure air temperature and vapor pressure. This last instrument consisted of dry-bulb and wet-bulb thermistors. The wet bulb was kept

FIGURE 2-10. Mirror Lake raft station showing wind speed, air temperature, and vapor pressure sensors. (Photo by Thomas Winter.)

moist by a wick extending into a reservoir of distilled water at the bottom of the unit. In 1992, the thermistor psychrometer was replaced by a Campbell Scientific model 207 probe, which measured air temperature with a thermistor and vapor pressure with a resistance-response circuit chip. This new instrument permitted collection of data during freezing conditions, which was not possible using the thermistor psychrometer (because the water reservoir and wick would freeze). In 1998, a Campbell Scientific CS500 probe replaced the 207 probe. In 1997, a shore station, with instruments suspended over the lake water surface, was established to replace the raft station. Both stations were operated simultaneously for two years to assure that the shore station would be a suitable substitute for the raft. Following the decision to use the Priestley-Taylor method to calculate evaporation (discussed later), wind data were no longer necessary. As a result, instruments at the shore station consisted of a thermistor to measure water surface temperature and a CS500 probe to measure air temperature and vapor pressure.

Output from the primary sensors was recorded by a digital data logger, which was programmed to scan the sensors every minute and calculate and store hourly and daily averages, as well as the maximum and minimum values and the minute they occurred for each day. Secondary analog instruments also were placed on the raft to provide backup data for wind speed and water temperature. In addition, an analog hygrothermograph was located near the lakeshore to provide backup data for air temperature and relative humidity near the lake.

The land station (Fig. 2-11) included a precision spectral pyranometer to measure incoming short-wave solar radiation and a precision infrared radiometer (pyrgeometer) to measure incoming long-wave atmospheric radiation. Data from these instruments also were recorded by a digital data logger programmed to generate data at the same frequency as previously described for the raft station. However, daily totals rather than averages were determined for radiation data.

In addition to the instruments located at the fixed stations, thermal surveys needed to be done at regular intervals to determine the heat stored in the lake. A thermal survey consists of taking water temperature measurements at discrete and consistent depth increments by lowering a submersible temperature probe from the surface to the bottom of the lake. Depth and temperature measurements were made at 10 locations in the lake, generally at weekly intervals. Temperatures were averaged for each specific depth increment "slice," and the heat stored in that slice was determined by multiplying the temperature by the heat capacity of water and the volume of water contained in each slice. The total heat stored in the lake was determined by summing the heat contained in all of the slices.

The Priestley-Taylor equation is a modified form of the energy budget, but it requires fewer data:

$$E = \alpha(s/s + \gamma)(Q_n - Q_x)/L \qquad (2\text{-}1)$$

where

E is evaporation from the lake surface (cm day^{-1}),

α is an empirically derived constant, usually 0.26, dimensionless,

$(s/s + \gamma)$ is derived from the slope (s) of the saturated vapor pressure-temperature curve at the mean air temperature, and the psychrometric constant (γ),

s was determined using an equation presented in Lowe (1977),

γ was determined using an equation presented in Fritschen and Gay (1979),

Q_n is net radiation (watts m^{-2}),

Q_x is the change in heat stored in the lake between thermal surveys (watts m^{-2}), and

L is the latent heat of vaporization (watts m^{-3}).

Evaporation rates were averaged over the days bounded by thermal surveys. Monthly evaporation was calculated as time-weighted daily values summed over the month, then multiplied by the area of the lake.

SURFACE WATER

Continuous records of discharge of the three tributary streams and the outlet were collected using Parshall flumes equipped with analog strip-chart recorders (Fig. 2-12). Widths of the flume throats were 30.5 cm for the NW Tributary flume, 23.9 cm for the W Tributary flume, and 7.6 cm for the NE Tributary flume. A factory-provided equation, calibrated to each flume, was used to convert stage measured at the inlet section of the flume to discharge.

FIGURE 2-11. Radiation sensors (on white post and on top of the white instrument shelter) at the land station next to the U.S. Forest Service Robert S. Pierce Ecosystem Laboratory. Precipitation and wind speed and direction sensors are also shown. (Photo by Thomas Winter.)

FIGURE 2-12. Parshall flume on the Northwest Tributary. (Photo by Thomas Winter.)

Surface water discharge from the lake was measured with a 61 cm wide Parshall flume located about 10 m downstream from the dam. The outlet flume was not installed until July 1990. Prior to flume installation, surface water discharge was determined by measuring lake stage and using a statistical relation between discharge measured at the flume and the stage of the lake (see chapter 4). We chose to measure outflow from the lake using a flume because we believed that the original method of gauging outflow using the dam as a broad-crested weir was not sufficiently accurate for our water budget purposes.

At very low flows, Parshall flumes can be somewhat inaccurate because the small amount of water passing through the throat is very shallow and it does not flow uniformly across the width of the throat. A V-notch weir plate was bolted to the downstream end of the outlet flume to increase accuracy at times of low flow (Fig. 2-13). This modification caused water to back up behind the plate to where the flume itself enclosed part of the storage pool. The same stilling well and float/recorder system used for the flume calculation also was used for the weir calculation.

FIGURE 2-13. Weir plate attached to the Parshall flume located downstream of the Mirror Lake outlet. (Photo by Thomas Winter.)

A notation was made on the flume chart to indicate when the weir plate was attached; during those times, the discharge equation for the weir was used instead of the discharge equation for the flume. The flumes on the three tributaries were not modified during low-flow conditions because their much narrower openings made it possible to measure low flows reasonably accurately.

GROUND WATER

The seepage of ground water to a lake and seepage of lake water to ground water are the most difficult components of a water budget to determine. Generally, it is a good policy to use, compare, and evaluate several different methods to determine acceptable values for the interaction of lakes with ground water. For this study, three methods were used to calculate fluxes of ground water to the lake and lake water to ground water: (1) analytical Darcy, (2) stable isotope ratios of oxygen, and (3) a numerical simulation model. The groundwater flux values presented in this chapter were determined as a compromise of the results of using all three methods. Pertinent details of the three calculation methods and how the groundwater flux values presented in this chapter were determined are presented in chapter 4. However, the field methods for construction and testing of groundwater wells and piezometers are presented next in this chapter.

Water table wells designed to determine the elevation of the top of the zone of saturation in the subsurface were drilled using a truck-mounted power auger. The holes were drilled to a depth of less than a meter below the water table, casing with a well screen attached at its base was lowered into the drilled hole, and sand was placed around the screen where a hole was drilled into glacial till. Where a hole was drilled into sand, the sand collapsed around the screen below the water table when the auger was removed. The annular space between the casing and drill-hole wall above the screen was back-filled with drill cuttings for both geological conditions.

Piezometers were used to measure hydraulic head at points within the groundwater system below the water table. Bedrock piezometers measured the hydraulic head of the ground water in the fractured bedrock beneath the glacial deposits. Piezometers in the glacial deposits measured the hydraulic head of the ground water at specific points within the deposits.

Bedrock piezometers were constructed by drilling a hole through the glacial deposits and 3 m into the bedrock using the mud-rotary method. Casing, with a drive shoe attached to its base, was lowered into the hole. Concrete was then pumped down the inside of the casing so it would flow out the end and fill the annular space between the outside of the casing and the drill-hole wall. Making use of the drive shoe, the casing was then pounded several centimeters into the bedrock. After allowing the cement to dry, the cement plug inside the casing was drilled out using the air-hammer drilling method, and drilling then proceeded to the final depth of the hole (Fig. 2-14a). Bedrock piezometers constructed in this way prevented any movement of ground water between the glacial deposits and the bedrock along the borehole wall. The open hole below the bottom of the casing allowed measurement of the average head of ground water in the bedrock fractures that intersected the borehole.

Piezometers in glacial deposits were constructed by drilling a hole to a desired depth using the mud-rotary method. An assembly made of a well screen attached to well casing, with a grout basket (also called a petal-cement basket) placed around the outside of the casing between the screen and the casing, was then lowered into the hole until the screen reached the bottom. Cement was pumped into the hole so it filled the annular space between the drill-hole wall and the casing in the interval above the grout basket. The grout basket kept the cement from entering the screened interval (Fig. 2-14b). By constructing piezometers in this way, the screened interval was isolated within the groundwater system, and the altitude of the water level in the piezometer was a measure of the hydraulic head at the point of the groundwater system where the screen was located.

Two or more wells completed at different depths at one location constitute a piezometer nest. In nearly all settings, each piezometer in the nest will have a different water level. This information is used to determine the vertical gradients within a groundwater system. For example, a deeper piezometer having a lower water level than a shallower piezometer indicates ground water is moving downward. Conversely, a deeper piezometer

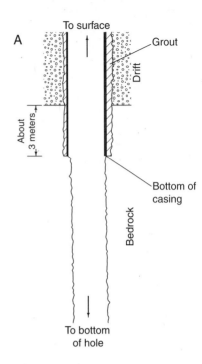

FIGURE 2-14. Diagram of (A) piezometer construction in bedrock and (B) piezometer construction in glacial deposits.

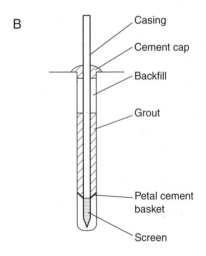

having a higher water level than a shallower piezometer indicates ground water is moving upward. Manual measurements of water levels in wells and piezometers were made weekly or biweekly using a steel or electric tape.

Hydraulic conductivity, K, which is a measure of how readily water moves through geological materials, is a parameter that needs to be measured or estimated in studies of ground water. At each well, K was determined by single-well hydraulic tests, also called "slug tests." Injection and withdrawal tests typically were conducted as follows: (1) a pre-test water level was measured in the well, (2) a solid object was placed in the well to displace a volume of water, causing a sudden rise in water level in the well, (3) frequent measurements of water level and associated time were made until the water level stabilized at the pre-test level, (4) the solid object was rapidly removed from the well, and (5) frequent water level measurements were made until the water level again stabilized. For some wells, where the recovery was very slow because K was small, only the injection test was conducted. Water level recovery curves were analyzed using a method described by Bouwer and Rice (1976) to determine K.

MEASUREMENT ACCURACY AND ERROR ANALYSIS

Every effort was made to measure each of the water budget components as accurately as possible. Our assumptions were that percent errors associated with determining each of the water budget components remained constant during the study period and that errors were randomly distributed. Our estimates of measurement error, based to a large extent on a thorough study of errors associated with water budget components of lakes (Winter 1981), are 5 percent for streamflow measurements, 5 percent for precipitation, 15 percent for evaporation, and 25 percent for groundwater flows to and from the lake. Error in determining change in storage in the lake was assumed to be 10 percent. (See chapter 4 for a detailed discussion of uncertainty of water budget terms.)

Cumulative errors were determined with a first-order error analysis, which assumes that each of the water budget terms is independent of the others. This obviously is a poor assumption. For example, all of the terms are dependent on precipitation. Nevertheless, a first-order error analysis usually provides a close approximation of the cumulative error associated with summing multiple measurements (Winter 1981). The equation for a first-order error analysis, assuming that errors are independent and randomly distributed, is

$$\delta = \sqrt{\delta_P^2 + \delta_E^2 + \delta_{SWI}^2 + \delta_{SWO}^2 + \delta_{GWI}^2 + \delta_{GWO}^2 + \delta_{\Delta L}^2} \qquad (2\text{-}3)$$

where

δ is error,

P is precipitation,

E is evaporation,

SWI is surface water flow to the lake,

SWO is surface water flow from the lake,

GWI is groundwater discharge to the lake,

GWO is loss of lake water to ground water, and

ΔL is change in volume of water contained in the lake (positive for increase in volume).

By measuring all of the water budget components of Mirror Lake, cumulative error (δ) can be compared with the residual (R) of the water budget equation:

$$R = P - E + SWI - SWO + GWI - GWO - \Delta L \qquad (2\text{-}4)$$

RESULTS

AVERAGE LAKE WATER BUDGET

The water budget components for Mirror Lake, averaged over the 20 years from 1981 through 2000, are shown in Fig. 2-15. Runoff provided by the three tributaries that discharge to Mirror Lake (SWI) was the largest input term; at 59 percent of all input terms, it was over twice as large as water supplied by direct precipitation (Table 2-1). Groundwater inflow (GWI) comprised 16 percent of water supplied to the lake. Loss of lake water to ground water (GWO) was the largest outflow component of the lake water budget, which is somewhat surprising for a lake that has a surface water outlet. The GWO averaged 51 percent of all the loss terms in the Mirror Lake water budget. It is likely that the dam is largely responsible for the large loss of lake water to ground water. By holding the lake at a higher elevation than normal, the lake area is larger, and the lakebed extends over highly permeable sediment along the south shore of the lake. Because the water table along this portion of the southern shoreline is lower than lake stage (Fig. 2-6), and because coarse sand and gravel are present south of the lake, large volumes of water can seep from the lake and flow as ground water to Hubbard Brook, about 0.3 km south of the lake (Rosenberry and Winter 1993; Rosenberry 2005). Evaporation provided 11 percent of the water losses from the lake, and it averaged 42 percent of annual precipitation during our study.

Substantial inter-annual variability of water budget components occurred during the 20 years as is evidenced by the large standard deviations for some of the water budget components (Table 2-1). Standard deviation as a percent of each water budget term ranged from 5 percent for *GWI* to 35 percent for *SWO*. Average annual change in lake storage was very small, but large variations existed from year to year, resulting in an average annual lake volume change of essentially zero (16 m³), as one would expect, but with a standard deviation of 2781 m³ (Table 2-1).

According to 20-year averages of each of the water budget components, water supplied to Mirror Lake slightly exceeded water lost from the lake (Fig. 2-15). That can't happen, of course, at least over the long term, but

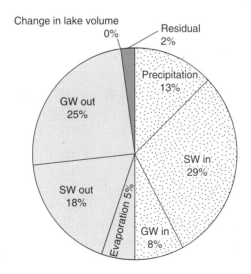

FIGURE 2-15. Mirror Lake water budget components averaged over the period 1981 through 2000.

given the relatively large uncertainties associated with the groundwater components in particular, an imbalance (referred to here as a residual term) of 2 percent of the water budget is remarkably small. Increasing *GWO* by 8.2 percent, well within the error estimate for ground water of 25 percent, would reduce the residual term to zero.

CLIMATE

How representative are the slices of the water budget pie for Mirror Lake relative to a longer-term perspective? To answer that question it is useful to compare the 20-year hydrologic record for Mirror Lake with longer-term precipitation records since, as mentioned previously, precipitation is the ultimate source for all of the water in Mirror Lake.

Precipitation at Mirror Lake averaged 1213 mm/year during the period 1981–2000 (Table 2-2). Longer-term records from the Hubbard Brook Ecosystem Study indicate precipitation was greater than normal during this 20-year period. At Hubbard Brook Experimental Forest Watershed 3, the instrumented watershed closest to Mirror Lake, precipitation averaged 1412 mm/year during 1981–2000 and was 7 percent larger than the long-term (1958–2001) annual average of 1320 mm/year (Bailey et al. 2003).

Data from the Palmer Hydrologic Drought Index (PHDI)(http://www1.ncdc.noaa.gov/pub/data/cirs/) provide a century-long perspective with regard to normal climate and decadal-scale deviations from normal. The PHDI is a monthly indicator of the degree to which hydrologic conditions are wetter or drier than normal. The PHDI is similar to the more widely used Palmer Drought Severity Index (PDSI)(Palmer 1965; Bailey et al. 2003) but has been modified to represent wetness and dryness better for hydrologic applications. The PHDI uses temperature, precipitation, latitude, and soil moisture retention data from National Weather Service climate stations within each National Weather Service climate region; a monthly index is calculated for each climate region. Values typically range from +7 to −7, being positive for wetter-than-normal and negative for drier-than-normal conditions. Descriptors are assigned to ranges of index values: normal (−0.5 to 0.5), incipient (±0.5 to ±1), mild (±1 to ±2), moderate (±2 to ±3), severe (±3 to ±4), and extreme (>+4, <−4).

Although Mirror Lake is in New Hampshire climate division 1, it is near the border between climate divisions 1 and 2, and variability in lake stage correlates better with PHDI data from climate division 2 than from climate division 1. Using climate division 2 as an indicator of long-term climate variability, hydrological conditions at Mirror Lake were wetter than normal during 1981–2000 (Fig. 2-16). The long-term (1900–2005) PHDI average is 0.03. The PHDI average during 1981–2000 is 0.87. Of the 240 monthly PHDI values during 1981–2000, 164, or 68 percent, were greater than 0.5. One monthly value, May 1984, was the wettest since 1902, and October 1995 through March 1999 was one of the wettest periods of the twentieth century (Fig. 2-16).

The PHDI data indicate that the climate region was substantially wetter than normal during 1981–1982, 1984, 1990, 1991–1992, and during all of 1996 and 1997 (Fig. 2-17). Numbered rectangles in Fig. 2-17 indicate the rank of the wetness or dryness of the water year (June 1 to May 31) relative to the 43 years from 1958 through 2001 as measured by rain gauges positioned in nearby Hubbard Brook Experimental Forest Watershed 3 (Bailey et al. 2003). Wet periods based on data from Watershed 3 correlate well with wet periods indicated by PHDI values. For example, the longest duration of PHDI values greater than +2 occurred during 1996 and 1997, and data from Watershed 3 indicate that those two water years were the second and ninth wettest of the 43 water years from 1958 through 2001 (Bailey et al. 2003). Mirror Lake stage is little influenced by these wet periods. Because the lake can spill over the dam when stage exceeds 213.28 m in elevation, stage rarely rises more than about 0.05 m above that spill elevation. The largest influence of wet periods is the maintenance of lake level at or above the spill-point elevation, which prevents the normal summertime stage decline. However, several exceptions exist (discussed in the next section), when a summertime lake stage decline either occurred during a wet period (e.g., 1996) or failed to occur during a near-normal period (e.g., 1986).

Dry periods were not as severe or as frequent during these 20 years. The PHDI data indicate only one period (1985) when values were substantially less than −2. Data from Watershed 3 indicate three water years that ranked in the top 10 driest of the 43-year period of record, one of which was the fourth driest (1984–1985 water year) and corresponds with the early portion of the 1985 drought indicated by PHDI values (Fig. 2-17). Declines in Mirror Lake stage were particularly large during several of these dry periods. The largest lake stage decline occurred during 1984, the fourth driest water year according to precipitation data

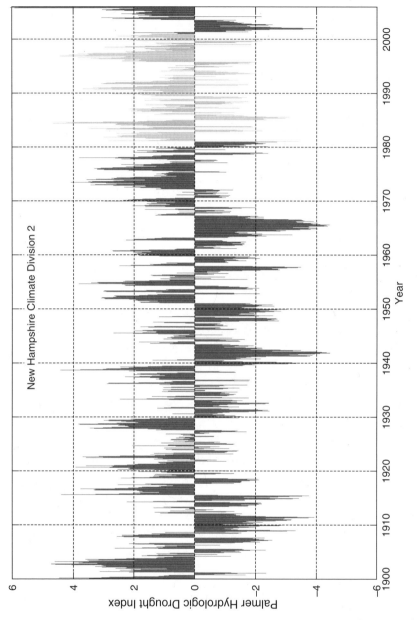

FIGURE 2-16. Monthly Palmer Hydrologic Drought Index (PHDI) values for 1900–2005. Study period is 1981–2000; values for most of the study period are shown in gray.

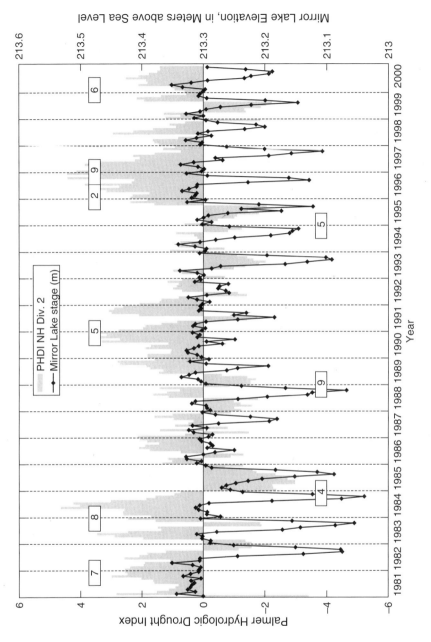

FIGURE 2-17. Palmer Hydrologic Drought Index and Mirror Lake stage during 1981–2000. Hubbard Brook Watershed 3 wetness or dryness rankings for 43 years (1958–2001) are shown in the boxes.

from Watershed 3. A large lake stage decline also occurred during 1988, the ninth driest water year based on Watershed 3 data. However, other large lake stage declines occurred during 1982 and 1983, years that were not particularly dry based on both PHDI and Watershed 3 data.

VARIABILITY OF THE LAKE WATER BUDGET

Although much can be learned by carefully measuring water budget components of a lake and comparing the relative magnitude of these components averaged on a decadal scale, a far better understanding of the hydrology of a lake can be achieved by measuring the response of various water budget components to changes in weather and climate averaged over shorter time scales. This section presents the Mirror Lake water budget from three perspectives: (1) interannual, (2) time-averaged seasonal, and (3) comparative, based on 240 monthly averaged periods.

Interannual Variability

Interannual variability in precipitation during 1981–2000 generated a range of responses of the other water budget components at Mirror Lake, depending on the timing, frequency, duration, and magnitude of the precipitation. Annual sums of the Mirror Lake water budget components indicate that SWI and SWO varied substantially from year to year, whereas GWI, E, and change in volume were the most stable (Fig. 2-18). In general, when annual precipitation increased, so did SWI and SWO. The three years having the greatest annual precipitation were 1981, 1990, and 1996. The three largest SWI values also were during 1981, 1990, and 1996. The fourth largest SWI occurred during 1984, a year when precipitation was only slightly above average. The largest residuals also occurred during the years having the greatest amount of surface water inputs and losses, indicating that even with a high level of accuracy in measuring streamflow, small errors of large terms result in relatively large overall errors (Fig. 2-18).

Mirror Lake stage typically declines during the summer months in response to reduced SWI and increased E (Table 2-2, Fig. 2-19). However, during four years of the 20-year study (1981, 1986, 1990, and 1992), Mirror Lake stage did not decline appreciably during summer (Fig. 2-17). The lack of stage decline is not surprising for the years

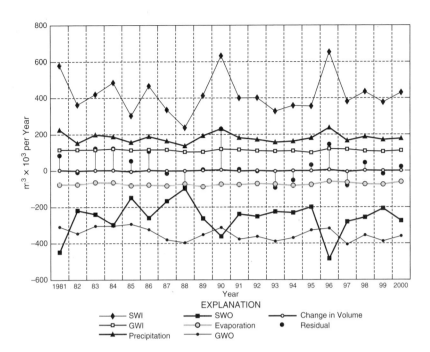

FIGURE 2-18. Annual components of Mirror Lake water budget including residual term. Positive values indicate additions of water to the lake. Negative values indicate losses of water from the lake.

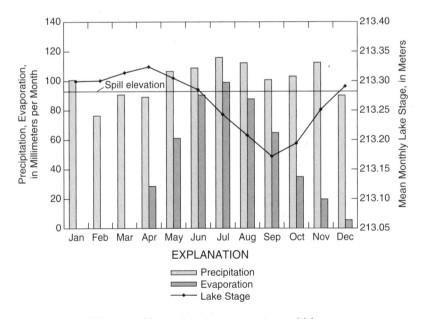

FIGURE 2-19. Mean monthly precipitation, evaporation, and lake stage.

1981 and 1990, when data from Mirror Lake (Fig. 2-18), Watershed 3, and the PHDI (Fig. 2-17) indicated that conditions were substantially wetter than normal. The reasons for the lack of lake stage decline during 1986 and 1992, however, are not so apparent. The amount of precipitation during both years was close to normal; 1986 precipitation was slightly above and 1992 precipitation was slightly below normal. Other years, such as 1983 and 1996 (Fig. 2-18), were substantially wetter, and yet lake stage declined during the summers of those years (Fig. 2-17).

Whether or not Mirror Lake stage declines during summer months is determined largely by the amount of rain that falls during those summer months and to a lesser extent by wet or dry conditions during the previous winter or spring. The stage of Mirror Lake, therefore, has a short hydrologic "memory"; lake stage responds primarily to hydrologic conditions during the most recent days to perhaps two months. As a result, an index such as the PHDI, which is designed to incorporate a substantial amount of antecedence, may not be an appropriate indicator for predicting Mirror Lake stage. A least-squares linear regression of total rainfall versus total change in lake stage during June through August explains 55 percent of the variance between those two terms. The two variables are plotted in Fig. 2-20 on separate y axes so the mean value for rainfall (50,800 m^3) is lined up vertically with the mean value for change in lake volume (−17,500 m^3). Change in lake volume is plotted as a surrogate for lake stage decline; the plots are scaled equally so that volumes can be compared between hydrologic budget components. In general, during years with large summertime rainfall (1981, 1990, 1998), the reduction in lake volume (or reduction in lake stage) is small; during years with small summertime rainfall (1983, 1985, 1999, 2000), the reduction in lake volume is large.

Deviations from this pattern are due to a variety of hydrological processes, although the size and frequency of individual rainfall events are among the most important factors. For example, summertime lake volume decrease was small during 1986, 1991, and 1992, and yet summertime precipitation was only slightly above average during those years. During both 1986 and 1991, lake stage was already declining at the beginning of June in response to a relatively dry May, and several rain events during August were large enough to raise lake stage 6 cm by the

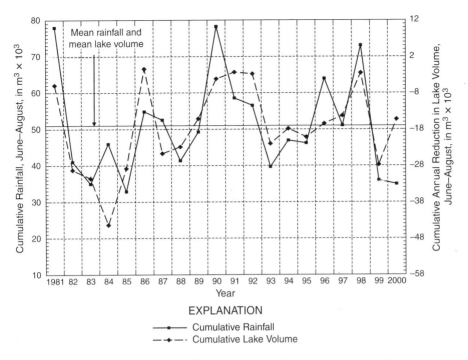

FIGURE 2-20. Cumulative rainfall and cumulative change in lake volume during June through August of each year.

end of August, leaving a relatively small net lake volume decline. During the summer of 1992, although no rain events were large, they were frequent and relatively evenly spaced throughout the summer, so lake stage decline was minimized. A similar situation occurred during 2000, when lake stage decline was reduced during a very dry summer by numerous small rain events during August. The opposite situation occurred during 1984, when slightly below-normal rainfall during summer resulted in the lowest lake stage of the 20-year study (Fig. 2-17). Rainfall was plentiful and slightly above normal during June and July, but during the last week of July through August almost no rain fell, resulting in a rapid and continuous lake stage decline during August that persisted through mid-October.

The inter-annual variability in the response of lake stage (and volume) to abundant precipitation followed by evaporative losses also occasionally depends on the timing and magnitude of specific precipitation events.

The summers of 1990, 1995, and 1999 provide examples during wet, normal, and dry years.

During July of 1990, lake stage fell relatively sharply because of only small to moderate amounts of rainfall that month (Fig. 2-21). Between August 6 and August 14, 28 cm of rain fell. This amount of rainfall caused the lake to rise rapidly to the second highest stage recorded during this 20-year study, which was well above the spill elevation. Very little rain fell during the remainder of the summer (until late September) and the lake stage quickly started falling because of losses to evaporation. However, the lake-stage minimum was not as low as it was before the rainfall, probably because losses to evaporation were being offset by water

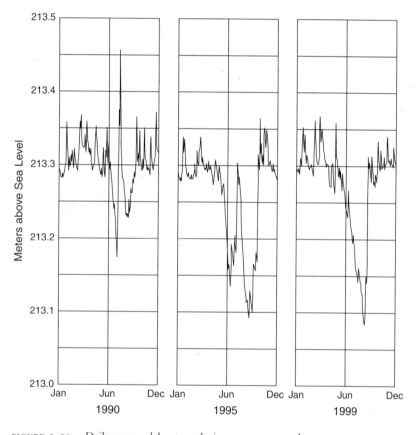

FIGURE 2-21. Daily average lake stage during 1990, 1995, and 1999.

draining from the saturated watershed of the lake by way of tributaries and ground water.

Although several scattered rainy periods occurred during July of 1995, evaporation rates were large, as they typically are during mid-summer, and the lake stage declined. From August 3 to 6, 9 cm of precipitation fell, resulting in a sharp rise in lake stage that briefly reached the spill elevation (Fig. 2-21). Very little rain fell during the remainder of the summer. Because a long evaporation season remained following the early August rainfall, the lake stage fell substantially until a steady rise began in early October.

During the summer of 1999, the stage of Mirror Lake declined in typical fashion. However, this year a substantial amount of rain fell during September. As a byproduct of hurricane Dennis, Mirror Lake received 5 cm of rain on September 10, and as a byproduct of hurricane Floyd, an additional 13 cm of rain fell on September 16–17. These inputs caused the lake stage to rise sharply from its summertime minimum to its spill elevation (Fig. 2-21). However, unlike 1990 and 1995, the evaporation season was largely over by mid-September, and the lake stage remained high for the remainder of the year.

Seasonal Variability

In spite of the substantial inter-annual variability in the hydrologic components, Mirror Lake stage maintains a relatively consistent seasonal variability that merits further discussion. The range of mean monthly stage was only 0.15 m (Fig. 2-19). Maximum mean monthly stage occurred during April in response to melting of the winter snowpack, and minimum mean monthly stage occurred during September in response to summertime evaporation. Mean monthly lake stage was higher than the spillway elevation during 7 of 12 months. Potentially high lake stages were essentially cut off because lake water spilled over the dam when the lake stage rose much above the 213.28 m elevation of the dam crest.

Precipitation was distributed relatively uniformly among the seasons (a condition common for the Hubbard Brook Valley [Likens and Bormann 1995]), which also aided in minimizing lake stage variability (Fig. 2-19, Table 2-2). The largest mean monthly precipitation occurred during July (117 mm) and November (113 mm), and the smallest mean monthly precipitation occurred during February (76 mm). Evaporation typically began during April, peaked during July, and then declined until the lake surface froze during November or December (Fig. 2-19).

Mean monthly precipitation and evaporation are overshadowed by several of the other hydrologic components of Mirror Lake's water budget (Fig. 2-22, Table 2-2). Surface water inputs from the three tributaries that discharge to the lake averaged 2.3 times more than precipitation directly on the lake on an annual basis, and the tributaries reached a peak during April when they were more than five times larger than precipitation directly on the lake. Precipitation exceeded surface water inflow to the lake during September, indicating that late-summer rainfall was retained in the watershed and that relatively little entered the lake via the tributaries. Mean monthly discharge of ground water to the lake was the most consistent of all the water budget terms, ranging from 8640 m^3 during February to 9880 m^3 during May.

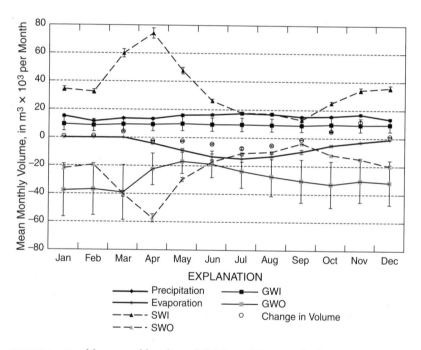

FIGURE 2-22. Mean monthly values of all Mirror Lake water budget components. Error bars depict fixed percentages from the mean values based on best estimates of measurement accuracy (*SWI, SWO*, Precipitation = 5%; Change in Volume = 10%; Evaporation = 15%; *GWI, GWO* = 25%).

Losses of lake water via surface water and ground water both exceeded loss due to evaporation during 9 of 12 months on a monthly mean basis (Fig. 2-22). During July through September, evaporation loss was greater than flow over the dam, but loss to ground water remained larger than evaporation. Flow from the lake via surface water was largest during April when mean lake stage was highest and smallest during September when mean lake stage was lowest. The plot of surface water outflow in Fig. 2-22 is a mirror image of surface water inflow but with a volumetric offset; surface water outflow values averaged 13,300 m^3 less than surface water inflow on a monthly basis. Mean monthly loss to ground water was greater than surface outflow during every month but April and May. During six months each year, mean monthly loss to ground water was the largest of all water budget components.

The seasonal pattern of flow from lake water to ground water (GWO) was different from the seasonal patterns of the other water budget terms because it depends on the relative elevations of both lake stage and adjacent ground water. During late winter, lake stage rose in response to increased inputs from the tributaries whereas infiltration to ground water had not yet increased. The hydraulic gradient therefore increased, and GWO typically was largest during March (Fig. 2-22). Soon thereafter, even though lake stage continued to rise as tributary inputs peaked during April (Fig. 2-19), infiltration to ground water from snowmelt increased and groundwater levels rose more than did lake stage. Therefore, gradients were reduced, and loss to ground water decreased substantially, typically from March to April. The smallest loss to ground water typically occurred during May (Fig. 2-22). Over the course of the summer, groundwater levels declined more rapidly than lake stage, and the rate of flow from lake water to ground water increased to a maximum value during October, after which recharge from late-season rains once again caused groundwater levels to rise more than lake stage, and loss to ground water was reduced slightly.

Change in lake volume also is plotted in Fig. 2-22 to show the change in the volume of water stored in the lake relative to the volumes of water added or subtracted from the lake. On average, the lake gained water during March, October, and November, and lost water during April through mid-September. The lake gained a slight amount of water during December, and there was virtually no change in volume during

January and February. April and September were transition months, during which the lake both gained and lost water so that the monthly average change in volume was small.

Error bars are presented in Fig. 2-22 to display the uncertainty associated with determining each term. Error bars also are included for the lake change in volume but are always smaller than 1500 m^3/month and are not visible because of the scale of the plot. Cumulative measurement error was determined through error analysis as described earlier and compared with the lake budget residual term. During 111 of 240 monthly water budget periods (46 percent), the water budget residual was smaller than the cumulative error (Fig. 2-23), providing an indication that estimates of measurement error were reasonable. Further analyses of hydrologic uncertainty are presented in chapter 4.

Month-to-Month Variability
Additional characteristics of the hydrology of Mirror Lake become evident when water budget terms are viewed on a monthly basis for the entire study (Fig. 2-24). One obvious feature is that Mirror Lake stage is fairly constant once the lake surface is high enough for water to spill over the top of the dam. Even very large additions of water from streamflow and precipitation result in small changes in lake stage. For example, streamflow both into and out of the lake exceeded 100,000 m^3 month^{-1} only four times during the 20 years (February 1981, April 1982, April 1994, April 1997), during which lake stage rise ranged from 2.7 to 4.6 cm. When lake stage was at or above the spill point, lake stage typically rose rapidly in response to precipitation events and then fell nearly as rapidly until the stage neared the spill point elevation (Fig. 2-25). As a result, the monthly mean values for lake stage track well with the larger-scale seasonal fluctuations in lake stage but do not represent well the short-term perturbations in lake stage, especially when the lake stage is high. Once below the spill elevation, lake stage is largely an integrated response to the cumulative balance between inputs and loss terms, and lake stage becomes much more sensitive to relatively small additions and removals of water. Monthly lake stage rises and declines were particularly large during July through September, but only when lake stage was lower than the 213.28 m spill elevation. Monthly lake

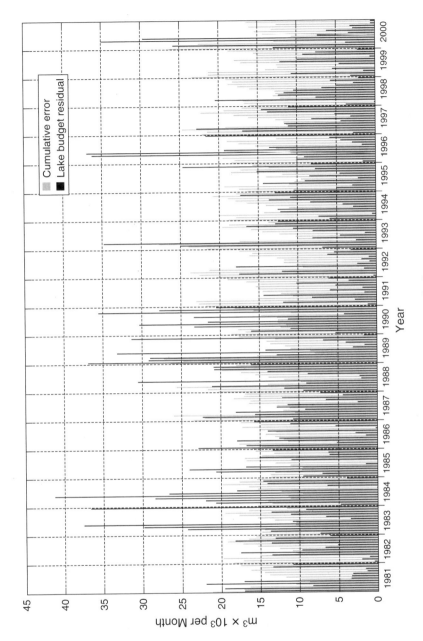

FIGURE 2-23. Cumulative error term compared to residual of water budget, 1981–2000.

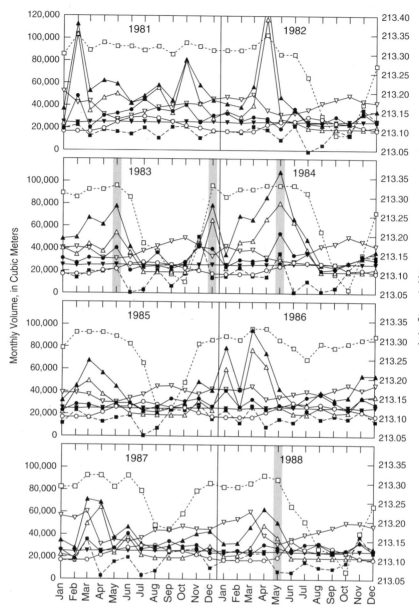

FIGURE 2-24. Lake water budget terms in m³ month⁻¹ and monthly lake stage in m above mean sea level. Shaded bars identify months when the water budget residual was larger than 30,000 m³.

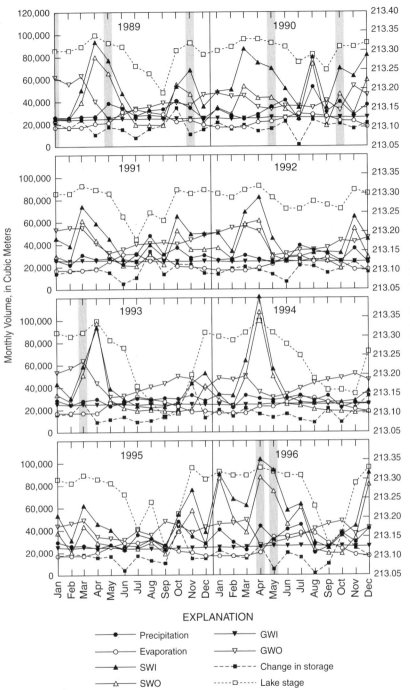

EXPLANATION

- ● — Precipitation
- ○ — Evaporation
- ▲ — SWI
- △ — SWO
- ▼ — GWI
- ▽ — GWO
- ---■--- Change in storage
- ---□--- Lake stage

FIGURE 2-24 (CONTINUED).

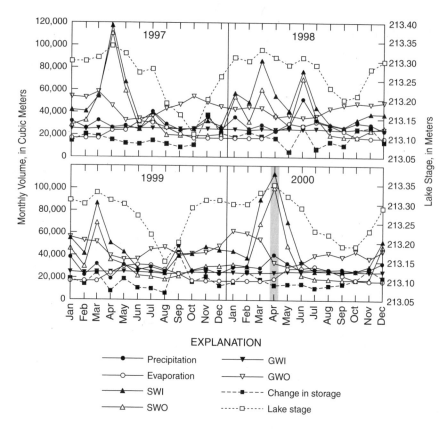

FIGURE 2-24 (CONTINUED).

stage changes of 5 to more than 10 cm during July through September were common. Monthly lake stage decline of 10 cm or more occurred 8 times during the 240-month study; monthly lake stage rise of 10 cm or more occurred 13 times during the study. All of these large changes in mean monthly lake stage occurred during late summer through early winter, and large lake stage rises ceased once lake stage reached the spill elevation.

With few exceptions, lake stage was maintained at or above the spill elevation as long as monthly precipitation was greater than monthly evaporation. For example, lake stage remained above the spill elevation during all of 1981 because rainfall was greater than evaporation during every month that year (Fig. 2-24). Once evaporation exceeded rainfall,

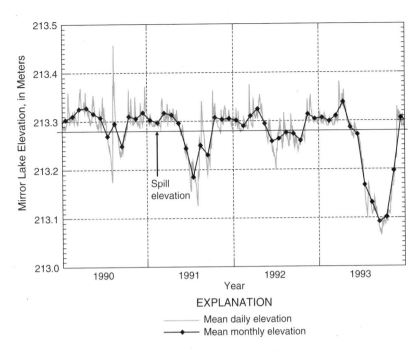

FIGURE 2-25. Daily compared to monthly lake stage during 1990–1993.

however, lake stage quickly declined to below the spill point and typically did not rise above the spill elevation until fall. Even large amounts of monthly rainfall during summer did not bring lake stage back to the spill elevation. For example, 28 cm of rain fell during August 1990, the second-largest monthly rainfall total of the study, and monthly average lake stage remained lower than the spill elevation (Fig. 2-24). Large rains during the early part of summer occasionally resulted in lake stage rising from below to above the dam crest, but those incidents were rare, occurring only during June 1987 and June 1998.

Water budget data also indicate that temperature during winter was important to the magnitude of the subsequent spring streamflow peak. During winters when streamflow was small and decreased monotonically, such as 1981–1982, 1988–1989, 1992–1993, 1993–1994, and 1999–2000, mid-winter rain on snow or mid-winter thaws of the snowpack likely did not remove much of the accumulated snowpack, and the subsequent streamflow rise during spring usually was quite large. However, when intermittent increases in winter streamflow indicated the likelihood of mid-winter thaws or rain on the snowpack, such as during 1983–1984, 1985–1986, 1989–1990, 1990–1991, 1994–1995, 1995–1996, 1997–1998, and 1998–1999, subsequent spring streamflow peaks were subdued relative to other years (Fig. 2-24). The mean of the largest monthly value for *SWI* during the five years when mid-winter thaws or rains likely were of minor consequence was 110,000 m^3. Streamflow peaked during April for all five years. The mean of the largest monthly value for *SWI* during the eight years when mid-winter thaws or rains on snow were evident was only 81,000 m^3. Spring streamflow following substantial mid-winter thaws peaked in March during six of those eight years, one month earlier than when mid-winter reductions in the snowpack were not as evident.

The mean residual in the monthly water budgets was 2500 m^3, and the standard deviation of the residual was 14,300 m^3. Several of the monthly lake budget residual values were quite large, exceeding 30,000 m^3 month^{-1} during 13 of the 240 months (Fig. 2-23). These 13 months are indicated with shaded rectangles in Fig. 2-24. A residual of 30,000 m^3 equates to a lake stage change of 20 cm, a substantial amount of water relative to typical lake stage changes measured at Mirror Lake. Nine of the 13 years when the largest water budget residuals occurred were during springs when streamflow was particularly large; one was during March, two were during April, and six were during May. The residual was positive, indicating a surplus of water, for 11 of the 13 months. Given the magnitude of these residual values relative to water budget components, only *SWI*, *SWO*, and *GWO* were large enough to include errors this large. Because *GWO* was quite stable from month to month, it is likely that the error resided in either the *SWI* or the *SWO* term. Because almost all of these largest residuals are positive, either *SWI* was smaller than measured or *SWO* was larger than measured. Eight of the 13 largest-residual months occurred prior to installation of the Parshall flume at the outlet of the lake, indicating that the equation used to calculate surface water outflow during the 1980s may have underestimated the largest outflows. This possibility, as well as sources of other measurement errors, is discussed further in chapter 4.

TABLE 2-1. *Twenty-year average of Mirror Lake annual water budgets*

Water budget component	Average, in m^3	Percent of total budget	Percent of inflow or outflow	Standard deviation, in m^3
P	181,981	13%	26%	26,403
SWI	417,485	29%	59%	105,315
GWI	112,541	8%	16%	5937
E	76,646	5%	11%	11,180
SWO	257,350	18%	38%	89,543
GWO	349,384	25%	51%	34,142
Change in lake volume	16	0%		2781
Residual	28,610	2%		83,475
Residual as % lake volume	3.3%			
Residual as % inflow	4.0%			
First order error	95,946			
Percent of budget	7%			

NOTE: Percentages not totaling 100 are due to round-off error.

TABLE 2-2. *Mean monthly lake stage and mean monthly sums of Mirror Lake water budget components, 1981 to 2000*

Mean monthly values, 1981–2000	Lake stage, in m above mean sea level	P mm/mo	E mm/mo	SWI mm/mo	SWO mm/mo	GWI mm/mo	GWO mm/mo
January	213.30	101	0	228	147	63	252
February	213.30	76	0	216	129	58	245
March	213.32	91	0	399	264	63	261
April	213.33	89	29	495	378	62	150
May	213.31	107	64	317	196	66	113
June	213.29	109	93	175	115	63	126
July	213.25	117	103	118	74	63	159
August	213.21	112	91	115	68	61	184
September	213.18	101	68	86	26	60	204
October	213.20	107	37	167	87	65	220
November	213.25	113	20	228	99	62	207
December	213.29	90	6	239	132	64	211
	Mean 213.268	1213	511	2783	1716	750	2329

NOTE: Values are given in mm of water applied to the 150,000 m² lake-surface area.

REFERENCES

Bailey, A.S., J.W. Hornbeck, J.L. Campbell, and C. Eagar. 2003. Hydrometeorological database for Hubbard Brook Experimental Forest: 1955–2000. General Technical Report NE-305, USDA Forest Service, Newton Square, Pennsylvania, 36 pp.

Bouwer, H., and R.C. Rice. 1976. A slug test for determining hydraulic conductivity of unconfined aquifers with completely or partially penetrating wells. *Water Resources Research* 12:423–428.

Buso, D.C., G.E. Likens, and J.S. Eaton. 2000. Chemistry of precipitation, stream water and lake water from the Hubbard Brook Ecosystem Study: A record of sampling protocols and analytical procedures. General Technical Report NE-275, USDA Forest Service, Newton Square, Pennsylvania, 52 pp.

Federer, C.A., L.D. Flynn, W.C. Martin, J.W. Hornbeck, and R.S. Pierce. 1990. Thirty years of hydrometeorologic data at the Hubbard Brook Experimental Forest, New Hampshire. General Technical Report NE-141, USDA Forest Service, Radnor, Pennsylvania, 44 pp.

Fritschen, L.J., and L.W. Gay. 1979. *Environmental Instrumentation*. Springer Advanced Texts in Life Sciences. New York: Springer-Verlag. 216 pp.

Likens, G.E., and F.H. Bormann. 1995. *Biogeochemistry of a Forested Ecosystem, Second Edition*. New York: Springer-Verlag. 159 pp.

Lowe, P.R. 1977. An approximating polynomial for the computation of saturation vapor pressure. *Journal of Applied Meteorology* 16(1):100–103.

Palmer, W.C. 1965. Meteorologic drought. Research Paper 45, U.S. Weather Bureau.

Rosenberry, D.O. 2005. Integrating seepage heterogeneity with the use of ganged seepage meters. *Limnology and Oceanography: Methods* 3:131–142.

Rosenberry, D.O., and T.C. Winter. 1993. The significance of fracture flow to the water balance of a lake in fractured crystalline rock terrain. XXIVth Congress International Association of Hydrogeologists-Hydrogeology of Hard Rocks. Geological Survey of Norway, Ås, Norway, pp. 967–977.

Rosenberry, D.O., T.C. Winter, D.C. Buso, and G.E. Likens. 2007. Comparison of 15 evaporation methods applied to a small mountain lake in the northeastern USA. *Journal of Hydrology* 340:149–166.

Shapiro, A.M., P.A. Hsieh, W.C. Burton, and G.J. Walsh. 2007. Integrated multiscale characterization of ground-water flow and chemical transport in fractured crystalline rock at the Mirror Lake Site, New Hampshire. In D.W. Hyndman, F.D. Day-Lewis, and K. Singha (eds.). *Subsurface Hydrology: Data Integration for Properties and Processes, 171st edition* (pp. 201–225). American Geophysical Union.

Tiedeman, C.R., D.J. Goode, and P.A. Hsieh. 1997. Numerical simulation of ground-water flow through glacial deposits and crystalline bedrock in the Mirror Lake area, Grafton County, New Hampshire. U.S. Geological Survey Professional Paper 1572, 50 pp.

Winter, T.C. 1981. Uncertainties in estimating the water balance of lakes. *Water Resources Bulletin* 17(1):82–115.

Winter, T.C. 1985. Mirror Lake and its watershed: A. Physiographic setting and geologic origin of Mirror Lake. In G.E. Likens (ed.). *An Ecosystem Approach to Aquatic Ecology: Mirror Lake and Its Environment* (pp. 40–53). New York: Springer-Verlag.

Winter, T.C., D.C. Buso, D.O. Rosenberry, G.E. Likens, A.M.J. Sturrock, and D.P. Mau. 2003. Evaporation determined by the energy budget method for Mirror Lake, New Hampshire. *Limnology and Oceanography* 48(3):995–1009.

3

NUTRIENT DYNAMICS

DONALD C. BUSO, GENE E. LIKENS, JAMES W.
LABAUGH, AND DARREN BADE

Long-term studies of the Hubbard Brook Experimental Forest watersheds near Mirror Lake have provided quantitative understanding about the importance of dissolved substance flux across ecosystem boundaries. Issues addressed by these well-known studies include the effect of regional atmospheric deposition; the effects of local watershed disturbances, such as clear cutting; and the change of natural ecosystem biogeochemical processes through time (Likens and Bormann 1995). The success of these efforts has been based on measurements of the volume and chemistry of precipitation and tributary water over more than four decades, and on the systematic derivation of ecosystem fluxes and chemical mass balances (budgets) from those data. The long-term record of the chemical fluxes for these Hubbard Brook Experimental Forest watersheds sets the stage for examining the response of Mirror Lake and its watershed and airshed to the same kinds of changes, and for quantifying comparable natural and disturbed ecosystem functions, using a similar approach.

The hydrologic setting of the upland, forested watersheds at the Hubbard Brook Experimental Forest is dominated by precipitation, tributary runoff, and evapotranspiration, with negligible groundwater flow (Likens and Bormann 1995). On the other hand, extensive studies of till depth and composition and of groundwater potentials around Mirror Lake confirm that ground water is a significant part of the hydrologic cycle in this lower-elevation watershed (chapter 2). Thus, our study of chemical fluxes at Mirror Lake, over the 20-year period of focus of this book, is particularly enhanced because of the direct and detailed accounting of all aspects of the water budget of the lake including evaporation and interaction of the lake with ground water.

Many natural and anthropogenic processes have played a role in determining the chemical characteristics of Mirror Lake. These characteristics are not static because the processes driving them are capable of change. The major ecological questions are as follows: how much do individual components of the system contribute to this dynamic nature, and how much of the change is due to natural versus anthropogenic factors? It may be relatively easy to quantify some sources, such as surface runoff chemistry or direct deposition of solutes via precipitation, but the contribution of groundwater inputs is often inadequately measured and even ignored in considerations of chemical characteristics of lakes. Long-term (on the order of millennia) climate change is clearly reflected in the chemistry of the lake sediments (Likens and Moeller 1985), whereas the effects of short-term (decades) climate change on solute retention or release have not been fully explored. Human activity within and outside of the Mirror Lake watershed for at least the past 200 years has also been a factor contributing to chemical change in Mirror Lake. Forest removal, construction of a mill dam, enhanced nutrient inputs from increasing numbers of homes, contamination by road deicers, and regional acidic deposition have all affected the fundamental characteristics of the lake to various degrees resulting in change in the chemical components of the water in the lake.

Many of these issues were considered in an earlier book about Mirror Lake (Likens 1985). The additional 20 years of record contained herein provide an opportunity to follow chemical and hydrological trends in the lake and its watersheds from a longer-term perspective, with complete and quantitative hydrologic data including ground water, and during a period of major external change (e.g., peaking and then declining inputs from acid rain). In the intervening decades since the publication of the earlier book, there has been growing awareness of the importance of placing studies in the context of climate change. The 20-year period we present here allows us to examine what that variability might be and what effect wetter or drier years might have on the chemical composition of Mirror Lake. Our analysis also is intended to provide perspective on the determination and use of water and chemical budgets for those limnologists who themselves will use such budgets to further the understanding of biogeochemical and ecological processes in lakes, and in particular relative to environmental problems.

The specific goals of this chapter are (1) to use water and chemical budgets of Mirror Lake to identify and better understand long-term trends in the chemical characteristics of Mirror Lake, (2) to quantify the role of ground water in those water and chemical budgets, (3) to identify the effect of wetter or drier years on those concentrations and budgets in the lake, and (4) to relate mass balances to long-term effects of disturbance and/or to change in natural ecosystem processes.

This chapter will focus on the variability and trends in chemical concentrations and fluxes at Mirror Lake during 1981–2000. The overarching issue in this chapter is the degree to which measuring chemical fluxes for all the hydrologic pathways is important in addressing the following questions:

- How has acidic deposition affected Mirror Lake, and can the response or recovery be quantified?
- How have various hydrologic pathways contributed to the contamination of Mirror Lake by road salt, and is that contamination continuing?
- Are there changes in nutrient concentrations and budgets within Mirror Lake, and what are the causes?
- Does ground water (in/out) matter to balancing chemical budgets for Mirror Lake?
- What influence do extremes in wet or dry years have on chemical budgets?

RESEARCH METHODS

During 1981–2000, the analytical procedures and instruments used to determine solute concentrations in samples collected at Mirror Lake changed little. Our diligence in maintaining stringent quality-control/quality-assurance protocols has helped to prevent many analytical problems, particularly with solutes in dilute systems like Mirror Lake, where solutes are often near the detection limit for the method of measurement. In addition, establishment of a routine sampling program at Mirror Lake in 1965 (Likens 1985) provided a stable prelude to the work described herein. Finally, the acquisition of personal computers in the 1980s and the application of a centralized data management system greatly improved data compilation and manipulation; although this change may seem trivial today, it has represented a substantial leap forward in efficiency since that time. A detailed discussion of each method, including analytical accuracy, precision, solute detection limits, and instrument upgrades, as well as quality control and database protocols are given by Buso et al. (2000). Procedures and precision of measurement are provided in Table 3-1.

In general, measurements of gases and pH were done as soon as possible in the nearby Robert S. Pierce Ecosystem Laboratory at the Hubbard Brook Experimental Forest. We performed the remaining solute analyses on samples shipped to Cornell University (1981–1983), and thereafter to the Institute of Ecosystem Studies. Analyses of dissolved organic and inorganic carbon (DOC and DIC, respectively) and acid neutralizing capacity (ANC) were not done routinely until 1991. The pH record prior to 1991 for the lake profile is incomplete because of gaps in the database. We chose not to reconstruct missing pH, ANC, DIC, or DOC data with modeled values. Only chemical data collected through an entire calendar year are included in this discussion. The lake profile, tributary, and outlet stream chemical data from 1967 to the present are available in original form at the Hubbard Brook Web site: www.hubbardbrook.org.

SAMPLE COLLECTION

Routine samples of precipitation, tributary, and lake water within the Mirror Lake watershed have been collected since 1963 as part of the Hubbard Brook Ecosystem Study (Likens and Bormann 1995). With few exceptions, the routine collection techniques used since the inception of the Hubbard Brook Ecosystem Study have not been modified. Like the analytical methods, these standard protocols and any substantial changes have been documented (Buso et al. 2000). Since the late 1970s, we have also collected samples of ground water from observation wells using a variety of methods depending on the physical and hydrologic characteristics of the site (Asbury 1990; Harte and Winter 1995; Winter 1985). The overview that follows discusses specific sampling challenges we have faced at Mirror Lake during 1981–2000 and our approach to solving those problems.

Lake Water

Profile samples of lake water are collected typically at least five times a year: (1) in January, in early winter after ice-in; (2) in March or April, in late winter before ice-out; (3) in April or May, in early spring after ice-out and lake turnover; (4) in August, in late summer at the time of highest surface water temperature; and (5) in October or November, in late fall after turnover. Weather conditions, personnel availability, and ice thickness have contributed to considerable variability in the actual date of sampling. Collections are always made from the deepest part of the lake (11 m maximum) about 100 m northeast from the physical center of the lake basin (Fig. 3-1). This site has a permanently moored buoy. Samples are normally collected at 0.5 m, 2 m, 4 m, 6 m, 8 m, and 10 m below the water surface, although this sequence varied more prior to 1990 when collections were done by many different colleagues.

Temperature and oxygen profiles are always measured with the chemical collections, and extra temperature and oxygen surveys are done in the months between the full chemical profiles to provide ancillary data on thermal structure and extent of mixing. Lake sampling methods and protocols are discussed in Buso et al. (2000) and by Likens (1985), and these follow the standard limnological procedures of Wetzel and Likens (1991).

The most important changes to our routine collection procedures during these 20 years were (1) filtration methods and (2) the use of a peristaltic pump. Water samples were filtered *ad hoc* before 1990 based on the requirements of specific studies. Beginning in 1990, a standard protocol was adopted, whereby all lake samples were filtered (0.45 μm–pore-size, GF/F grade Whatman® paper). Second, before 1990, various brass, glass, and plastic collection vessels (e.g., Van Dorn bottle) were used to lift samples from discreet depths of the lake. A peristaltic pump with weighted Tygon® tubing has been used since then, and this facilitates efficient, in-line filtration.

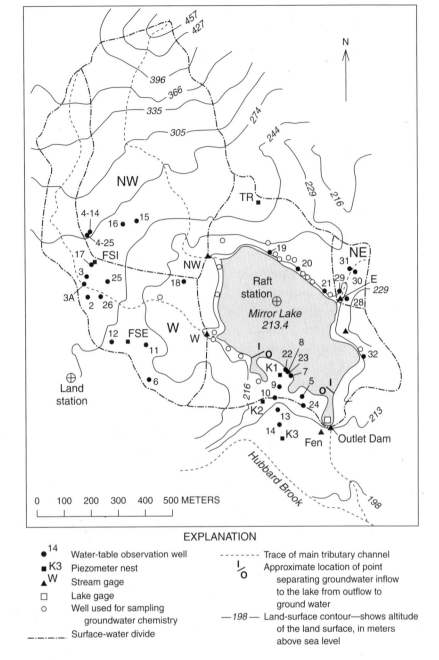

FIGURE 3-1. Locations of wells used to sample ground water. Locations of hydrologic instruments as in Fig. 2-7 are also shown.

Precipitation
Samples of precipitation were collected for chemical analysis at a single site, next to the Robert S. Pierce Ecosystems Laboratory, about 0.4 km west and 40 m higher than Mirror Lake (Fig. 3-1). This site, which also contains a gauge for measuring precipitation volume along with other meteorological instruments, has been in continuous operation since 1963. We use bulk precipitation collectors: that is, the 25 cm diameter rain funnel (or 21 liter snow bucket) is open to the atmosphere for the entire period of collection, which is approximately one week (normally 6–8 days)(Buso et al. 2000). The chemical inputs to such collectors are construed to be a combination of wet-only solute deposition and dry particulate deposition, but the dry deposition and any gaseous inputs are not quantitative (Likens and Bormann 1995). Another gauge for precipitation volume, about 0.5 km east of Mirror Lake, does not have an associated collector for chemical analysis. Variability in precipitation volume between standard rain gauges at Hubbard Brook Experimental Forest is on the order of 10 percent (Bailey et al. 2003), but the chemistry of bulk collectors arrayed across the valley is nearly identical on a weekly basis (Likens, Buso, and Hornbeck 2002). Based on these data, we determined that an additional chemical collector at Mirror Lake was unnecessary. However, year-round weekly collections of precipitation are necessary, because temporal variability in precipitation chemistry can be several orders of magnitude greater than spatial variability at Hubbard Brook Experimental Forest (Likens and Bormann 1995; Buso et al. 2000).

Surface Water
Samples of tributary inflow or outflow are taken from a position as close to the lake as possible (Buso et al. 2000). At the outflow dam on the lake, the samples are taken from the flow over or from leakage through (in the case of low lake surface level) the spillway boards. At each of the three tributaries, the samples are taken at the last flowing waterfall above the still water of the lake (Fig. 3-1). Ancillary chemical data suggest that, in the Mirror Lake catchment, these seemingly insignificant choices are actually important. For example, during low-flow conditions, samples taken from below the spillway on the bedrock lobe on which the dam is placed have much larger pH, ANC, and DIC than those collected

from a few meters above on the lake spillway itself. The lower site may incorporate rock-fracture flow that is not part of the lake hydrologic balance, or some deep seepage that has been chemically altered after leaving the lake through the sediments. For the purposes of monitoring lake outputs, the upper spillway site was selected as the better choice for measuring actual lake outflow chemistry.

Collections from a site on the Northeast Tributary, 5 m downstream toward the lake from the routine sample point, yielded consistently greater pH, ANC, and DIC values, possibly due to seepage water forced to the surface near the lake and water table interface. In this case, the routine site was moved downstream to capture more accurately all the inputs to the lake. In this way, the Mirror Lake watershed is distinctly different from the shallow flow-path watersheds at Hubbard Brook Experimental Forest, where changes of 20 m or less in the site for sample collection are rarely detected in the water chemistry (Buso et al. 2000; Likens and Buso 2006). Tributary samples are not filtered, except for DOC, because turbidity is normally very low in these waters (Buso et al. 2000).

Incoming Ground Water

The samples of incoming ground water (GWI) taken for this study are from 39 different water table wells within the Mirror Lake watershed that are upgradient from the lake. The wells are in that part of the groundwater flow system that flows to Mirror Lake (Fig. 3-1). Water table wells have a permeable screen located within the upper portion of the saturated groundwater level. Wells were installed starting in 1978. Details on construction of the water table wells are found in chapter 2 and in Winter (1984). The 20 water table wells located in ground water flowing to the lake, from which water was withdrawn for chemical analysis through 2000, commonly were some tens or hundreds of meters from the lake edge. In 2004, an additional 19 water table wells were installed in close proximity to the lake edge (commonly within 10 m). Chemical analyses from water withdrawn from those wells in 2004 were used with data obtained between 1981 and 2000 in the calculations described below. There is little evidence to suggest the local crystalline bedrock–fracture flow system contributes much ground water to Mirror Lake (chapter 2), so measuring the chemical characteristics of water flowing through the glacial overburden (till) was the focus of our efforts. Nevertheless, samples were also collected and analyzed from numerous bedrock wells, so the composition of local ground water in the crystalline rock is known.

The frequency of collection varied greatly among wells; some were sampled only a few times in 20 years, and some were sampled on dozens of dates. Collections of ground water were made with a variety of externally mounted peristaltic pumps where water levels were approximately 8 m deep or less. In wells with water levels deeper than about 8 m, bladder-type or centrifugal pumps were lowered into the well to push water to the surface. The standard procedure was to purge all water in the well pipe three times and then to collect a sample when the level returned to pre-pump depths after the third purge. All groundwater samples were filtered in-line and on site, using 0.45 μm–pore-size, GF/F grade Whatman® paper. The groundwater pH values reported here are from air-equilibrated samples, measured in the Robert S. Pierce Ecosystem Laboratory, not from field meters.

Outflowing Ground Water

Solute concentrations in groundwater outflow (*GWO*) were operationally defined as the solute values of lake water within the zone of out-seepage, *before* entering the lake sediments. It was not necessary to account for changes in solute chemistry within the sediments (e.g., increases in base cations, pH, nitrate, ANC) once the water left the lake. Geophysical studies of the out-seepage zone on the south side of the lake found the most conductive material was in a narrow band of sand and gravel sediments from 0 to 2 m deep (Asbury 1990; Winter 1985). Organic-rich material in the lake sediments deeper than 3 m greatly reduced the hydraulic conductivity of the sediments. Solute concentrations in the outlet stream and the shallow lake water were typically identical, with minor differences only during brief, high-flow events under ice-covered conditions. The weekly outlet stream chemical analyses were used to calculate out-seepage chemical flux, rather than using the 0 to 2 m lake profile samples, which were collected only six times a year. The out-seepage flow data were compiled on a monthly basis (chapter 2), so the *volume-weighted monthly average* outlet concentration data were multiplied by the measured monthly out-seepage flow to calculate monthly solute outflux. As a result of these decisions, the annual volume-weighted average concentrations of solutes in *GWO* are slightly different from the annual average concentrations both in the lake and the outlet stream (Table 3-2).

SPECIFIC ANALYTICAL PROBLEMS

Changes and problems with analytical procedures appear to have affected the long-term trends for ammonium and phosphate.

Ammonium Measurements

Other than in anaerobic hypolimnetic samples, ammonium was typically found in very low concentrations; 31 percent of the 6700 ammonium values were below the method detection limit of 0.01 mg L^{-1}. Thus, any change in the ammonium measurement technique could have had a large effect on the chemical record. The ammonium measurement was switched in 1993 to a new continuous-flow analyzer with essentially the same colorimetric method as before, and the two instruments were run in parallel for several months, which is our standard procedure (Buso et al. 2000). No significant differences between the techniques were detected at that time using the full array of samples we collect (a non-parametric rank sum test was applied to compare results because these data were skewed toward very low values [i.e., not normally distributed]). However, by observing the long-term trends at each site in retrospect, it was found that the average ammonium concentrations evidently decreased slightly, though abruptly, after 1992. This change may have been in response to improved background color correction, because it was only observed in colored samples (relatively humic sites), specifically the Northeast and West Tributaries to Mirror Lake, but in no other samples of tributary water, lake water, or precipitation. Although a change in ammonium could have been due to human influence, because the West Tributary watershed contains several seasonal homes, a step decline in ammonium was also seen in the Northeast Tributary, which has no such development. The observed ammonium decline was from 0.04 to 0.01 mg L^{-1} and from 0.03 to 0.01 mg L^{-1}, on average, for the Northeast and West Tributaries, respectively. Because no further concurrent testing of instruments could be attempted, we conclude that the ammonium concentration data from the Northeast and West Tributaries to Mirror Lake prior to 1993 must be used with caution.

Phosphate Measurements

A second problem involved the phosphate measurements. Phosphate was also found only at very low concentrations; 44 percent of the 6700 phosphate values were below the method detection limit of 5 μg L^{-1}.

The colorimeter in the continuous flow instrument used to measure phosphate failed suddenly in 1988. Thus, no parallel analyses could be performed to determine a correction factor between the instruments. From a review of the long-term phosphate concentration data, we observed a gradual and unexplained phosphate increase (from an average of about 10 µg L^{-1} to an average of 19 µg L^{-1}), along with much greater variability in samples of tributary water, lake water, and precipitation, beginning about 1986. A relatively large drop both in phosphate concentrations and variability for those sites followed in 1989 with the upgraded colorimeter (concentrations were reduced by an average 10 µg L^{-1} for all sites). We conclude that the phosphate concentrations from the period 1986 through 1988 are problematic, and conclusions drawn solely from the concentration data during that period should be made with caution. However, because the problem affected all samples of precipitation, tributary water, and lake water with the same offset value, the ratios between the various mass inputs and outputs to the lake should not be affected. The instrument was replaced in 1992 with an extensive period of parallel measurements made between the new and old analyzers, and no correction for the last eight years of the study are necessary.

DATA CALCULATIONS AND ASSUMPTIONS

Lake Chemical Mass Changes

Lake mass was calculated for each time samples from the profile were collected and analyzed, typically four to six times each year. The volume (in m³) of each 1 m stratum in which a sample was taken was multiplied by the concentration (mg L^{-1}, µeq L^{-1}, µmol L^{-1}) in the stratum to obtain a mass for each layer (in kg, eq, or moles). Volumes were derived from previous bathymetric surveys (Likens 1985, p. 50). For the strata between discreet chemical samples, an average of the concentration in the adjacent layers was used. We collected temperature profile data more frequently than the chemistry samples, and at intervals as small as 0.5 m through the water column. These ancillary temperature data suggested that a gradual, linear extension between sample points more accurately described chemical changes with depth than a model with abrupt, step-wise changes.

One problem intrinsic to water and chemical budgets in lakes within the northern Temperate Zone is the effect of ice cover and accumulating snow on quantifying nutrient pools and inputs. The chemistry of the ice pack was ignored in our annual calculations. Also for continuity with other fluxes, the frozen precipitation that built up on the ice was added into the annual budget during the month in which it fell, rather than at the time of ice melt in April each year (see chapter 5 for a full discussion of the ramifications).

Because the profile samples were not collected on the same date each year, a simple process was developed to produce an annual value from which a change in lake storage could be calculated based on a fixed date. The solute masses for each stratum for each sample date were totaled to obtain a whole-lake value, and this mass was divided by the known lake volume to create a volume-weighted average (VWA) concentration for the lake. These whole-lake concentration data were plotted in time-series graphs to allow for visual estimation of VWA concentrations on 31 December for every year, from 1980 to 2000. This interpolated concentration value was multiplied by the measured lake volume on 31 December to obtain a lake mass for each nutrient at the end of each year. The mass at the start of a year (M_1) was subtracted from the mass at the end of each year (M_2) to estimate annual change in storage (i.e., $M_2 - M_1 = \Delta S$). By definition, positive ΔS values indicated an increase in solute mass within the lake during a year, whereas negative ΔS values indicated a decrease. Seasonal changes in ΔS, which could be significant for some solutes, did not affect these annual values.

All chemical inputs and outputs to the lake were expressed as mass-per-unit time, not as mass-per-unit area-time, as is the case for the long-term watershed comparison work at Hubbard Brook Experimental Forest (Likens and Bormann 1995). This approach was used because Mirror Lake was the central accounting unit into/from which each flux was measured. However, for the purposes of comparing retention of atmospherically derived nutrients among the sub-watersheds, the inputs were normalized to each area, and those units were calculated and stated explicitly as flux per hectare-time (see Buso et al. 2000).

Input Estimates

Daily chemical inputs via atmospheric deposition, also referred to herein as precipitation (in kg, eq, or moles), were calculated by multiplying the weekly chemical concentrations (in mg L^{-1}, µeq L^{-1}, or µmol L^{-1}) by the daily volume of precipitation (in m^3) for that weekly period. The precipitation loading was derived from the average volume of the two rain gauges bracketing the lake's watershed and was based on a lake surface area of 15 ha; no attempt was made to account for small changes in lake area due to variation in lake water level and volume.

Daily streamwater inputs via the three tributaries were calculated by multiplying the weekly chemical analyses and the daily streamwater volumes (same units as precipitation) to produce a daily flux for each discreet sample date. For days between discreet samples, when no actual streamwater sample was collected, the average chemistry of the previous and next streamwater sample was applied to the daily water volume to create a daily flux value. This procedure is essentially the same as that used for the Hubbard Brook Ecosystem Study for the past 40 years (Buso et al. 2000).

The daily fluxes for precipitation and tributary water were compiled into monthly and annual calendar-year fluxes. Monthly or annual volume-weighted concentrations were calculated by dividing total mass flux by total water volume for the period in question.

Groundwater influxes were based on measured monthly inflow volumes (chapter 2), multiplied by the median chemical values determined from the samples collected from water table wells. Medians were chosen as representative of the groundwater flow system to avoid the influence of extreme values (low or high concentrations) in the calculation of averages. With the use of a fixed median value for incoming groundwater concentrations, variability in groundwater influx is due exclusively to changes in hydrologic flux. Ground water was sampled much less often than the other chemical collection sites from 1978 to 2000. In addition, most of the groundwater wells were not in close proximity to the shore of the lake. The wells from which water was withdrawn for chemical analyses in 2004 expanded the spatial extent of the flow system for chemical characterization. Thus, we included analyses from wells sampled in 2004 in our characterization of the chemical characteristics of ground water flowing into the lake. Data from the most frequently tested wells indicated that the chemical characteristics in water from individual wells changed little over time, a result consistent with the findings of other studies of chemical inputs to lakes from ground water (LaBaugh 1991; Krabbenhoft and Webster 1995; Wentz et al. 1995).

Wells in the watershed of Mirror Lake are not uniform in chemical composition. These differences result from the three-dimensional characteristics of ground water flowing into Mirror Lake, where inputs consist of shallow, shorter flow paths, intermediate-depth flow paths, longer flow paths, and deeper, much longer flow paths (Winter 1981). In general, the higher concentrations in ground water represent water that has been in contact with geologic material in the subsurface for some time and that commonly has longer flow paths. Alternatively, wells that are in areas where ground water is recharged by precipitation, whereby the water moves rapidly through the soil and till, have concentrations lower than elsewhere in the groundwater flow system. These short flow paths may be in close proximity to the lake and the streams. Furthermore, some wells close to the lake shore can be diluted by input from lake water due to seasonal reversals in direction of flow (chapter 2) so that concentrations are lower than elsewhere in the flow field. The relative contributions of these flow paths to overall input from ground water flowing into the lake are not easily determined. Median concentration is a robust representation of the central tendency of the chemistry of ground water flowing into the lake and is what we used.

Output Estimates

Daily chemical output from Mirror Lake was calculated using the weekly chemistry of the outlet stream in the same manner as the three tributaries and was compiled into monthly and annual values in the same way. Groundwater chemical output (i.e., out-seepage) was also calculated using the weekly outlet stream chemical data aggregated to produce a monthly volume-weighted average concentration. Outlet stream chemistry is a proxy of near-surface concentrations in the lake, and the frequency of sample collection at the outlet was greater than sample collection within the lake. Monthly volume-weighted average outlet concentrations were multiplied by the measured, monthly groundwater outflow to provide monthly groundwater nutrient out-seepage and were then compiled into annual data.

Statistical Analyses

Statistical analyses were performed using SigmaStat® 3.5 software and the Statistical Analysis System (SAS). Simple linear regression models were used to determine whether the variability in annual volume-weighted concentrations or annual fluxes could be accounted for by year. In the case where a statistically significant relation was revealed, the results of the analysis are reported in terms of the regression coefficient and the probability of a greater F statistic resulting from chance that concentration (or flux) and year were related. Unless otherwise stated, the number of observations used in the annual regression analysis was 20. The confidence level of the test was $\alpha = 0.05$. The Shapiro-Wilk test was used to determine if data were normally distributed. The confidence level was $\alpha = 0.1$. If data were not normally distributed, then relations between parameters were determined using the Kendall tau test. Results of the test are reported as the Kendall tau correlation coefficient, and the probability of a greater absolute value of the correlation coefficient resulting from chance. The Pearson's r test was used to determine if two parameters, both representative of a normal distribution, were significantly correlated. The confidence level for the test of a significant relation for both the Kendall tau and Pearson's correlations was 95 percent.

RESEARCH RESULTS

CHEMICAL CHARACTERISTICS OF WATER SOURCES

Quality Control Results

The volume-weighted composite average ion balances of equivalents for the lake samples suggested a good match between cations and anions (Table 3-2), given a reasonable charge assigned to DOC (Buso et al. 2000). Average ion balances and percent ion error for volume-weighted composite precipitation samples were low (5 µeq L^{-1} imbalance and less than 5 percent error (Table 3-2), indicating that most solutes were accounted for and measured with reasonable accuracy. Average ion imbalance and percent ion error for a 20-year volume-weighted composite of the three tributaries combined were 3 µeq L^{-1} and 1 percent, respectively (Table 3-2). The ion balance for individual groundwater samples commonly was within 5 percent, although a few samples of well water had evidence of dissolved iron (rusty precipitate) and had larger imbalances, perhaps indicating the presence of unmeasured iron. Examination of median values of inflowing ground water indicated the sum of anions exceeded the sum of cations by 53.6 µeq L^{-1} (Table 3-2). The difference is primarily a function of the fact that the median concentration was used for each ion, so a balance of the resulting sum of anions and cations was not expected.

Mirror Lake Chemistry
Based on concentration expressed in equivalents, the base cations calcium, magnesium, sodium, and potassium represented 99 percent of the cations and constituted approximately 50 percent of the total ionic charge (Fig. 3-2). Sulfate, chloride, and bicarbonate represented 96 percent of the anions and constituted 48 percent of the total charge. On the basis of the annual volume-weighted average concentration in equivalents for the years 1981 to 2000, the most abundant cation and anion were calcium and sulfate. However, the annual average sulfate concentration value, expressed in $\mu eq\ L^{-1}$, was only slightly greater than the chloride and bicarbonate values (Table 3-2). During the period of observation that is the focus of this book, the relative abundance of the major ions based on concentrations in equivalents changed substantially (Fig. 3-3) (see chapter 6). In 1981, calcium and sulfate were the most abundant cation and anion, representing some 24 percent and 21 percent of the ionic charge, respectively. As calcium and sulfate concentrations declined over the 20-year period, sodium and chloride increased. Chloride became the most abundant anion by 1993, and sodium was similar to calcium by 2000. Thus, the decline in calcium and sulfate ions was matched by the increase in sodium and chloride ions so that the total ionic concentration did not decline in the period 1981 through 2000. The 20-year average value for total ionic concentration in the lake was 538 $\mu eq\ L^{-1}$, with a standard deviation of 15 $\mu eq\ L^{-1}$, and a minimum of 516 in 1990 and a maximum of 561 in 1995 (Fig. 3-3).

Bulk Precipitation Chemistry
The chemical characteristics of bulk precipitation at Mirror Lake from 1981 through 2000 were dominated by acidic deposition (Tables 3-2 and 3-3), and this is clearly shown by the average precipitation pH of 4.3. The average pH was determined by calculating the average hydrogen ion concentration and then converting that concentration to pH. The most abundant cation and anion were H^+ and SO_4^{2-}. The pH in precipitation was 156 times more acidic than the average pH of 6.5 for lake water. Ionic strength of average precipitation was about 30 percent of that for the lake. The solutes that represented most of the ionic composition of precipitation were hydrogen ion (which made up about 33% of the total ionic charge) and ammonium (8%), and the anions sulfate (27%) and nitrate

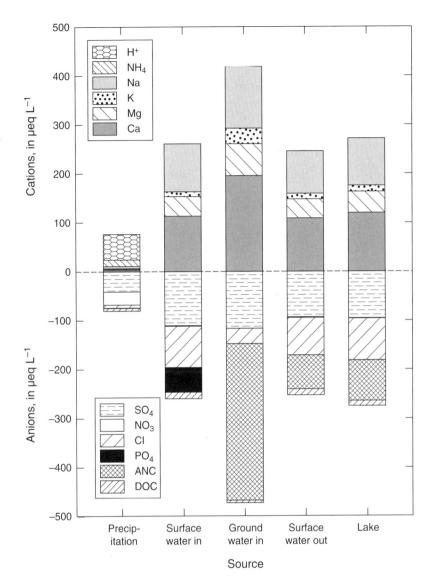

FIGURE 3-2. Average ionic charge distribution for bulk precipitation, composite (volume-weighted) tributary water in, ground water in (based on median data), surface water out, and the lake profile.

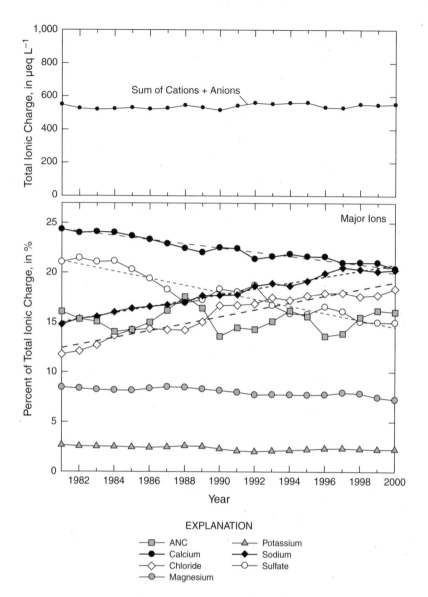

FIGURE 3-3. Annual average total ionic charge over time, and average charge distribution trends (dashed lines) in the lake for the major ions (>1% of total charge).

(17%)(Fig. 3-2). Annual average concentrations of calcium, magnesium, sodium, and dissolved silica in precipitation were small. Generally, the annual concentrations were relatively uniform over time. The exceptions were ammonium, hydrogen ion, nitrate, and phosphate, which were highly variable, exhibiting up to twofold differences between years. These characteristics are typical of the regional precipitation chemistry for the period 1981–2000 and are essentially identical to the variations in precipitation chemistry collected at nearby Hubbard Brook Experimental Forest during this period (Likens, Buso, and Hornbeck 2002).

Tributary Water Chemistry
Each of the three tributaries, West, Northwest, and Northeast, has its own individual chemical characteristics (Table 3-2; Fig. 3-4a, b, and c). As noted in chapter 2, the three tributaries represent 29 percent of the average annual total water input between 1981 and 2000, with the Northwest Tributary contributing the most water of the three, and the Northeast the least. Based on annual average concentrations, the West Tributary was the most similar of all the tributaries to the lake in relation to concentration of individual ions and the changing relation of those ions over time. Calcium and sulfate were the most abundant cation and anion in the West Tributary in 1981. These ions declined in concentration with time, and at the same time, concentration of sodium and chloride increased, such that in 2000, sodium and chloride were the most abundant cation and anion in the West Tributary based on annual volume-weighted average concentration. Calcium and sulfate were the most abundant cation and anion in the Northwest Tributary, and these declined in general over the 20 years of study. Over the same period in the Northwest Tributary, sodium and chloride increased, yet these ions were still present in smaller concentrations than were calcium or sulfate in 2000. The most abundant cation and anion by far in the Northeast Tributary were sodium and chloride. A general increase in the concentration of these ions was evident from 1981 to 1995, followed by a period of somewhat similar values smaller than the peak in 1995 but larger than in 1981.

The chemical characteristics of the three tributaries combined, expressed as an annual volume-weighted average concentration for total tributary input, are similar to that of the lake (Table 3-2; Fig. 3-3). Based on concentration expressed in equivalents, the base cations calcium,

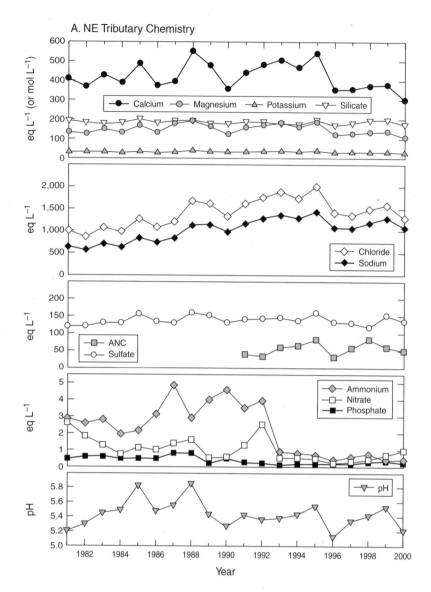

FIGURE 3-4. Annual volume-weighted average concentration time trends in the (A) Northeast, (B) Northwest, and (C) West Tributaries to Mirror Lake. Units are in µeq L^{-1}, except for dissolved silicate (µmol L^{-1}) and pH.

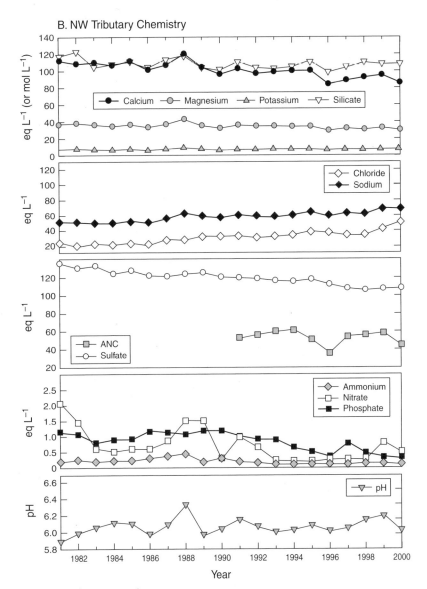

FIGURE 3-4 (CONTINUED).

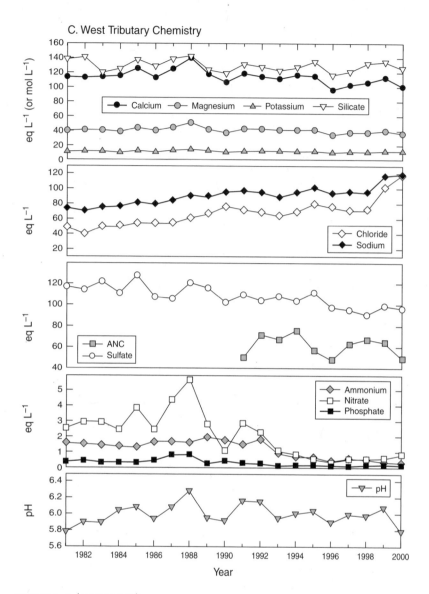

FIGURE 3-4 (CONTINUED).

magnesium, sodium, and potassium represented 99 percent of all cations and constituted 50 percent of the total charge in the combined tributary input. Sulfate, chloride, and bicarbonate represented 47 percent of the total charge and 94 percent of the anions from all the tributaries flowing into the lake (Table 3-2). The decline in calcium and sulfate concentrations over time, coincident with increase in the sodium and chloride concentrations, resulted in sodium and chloride replacing calcium and sulfate as the most abundant cation and anion in the combined input of tributaries to Mirror Lake (Fig. 3-5). One distinct difference from the lake shown by all the inflowing tributaries was that the annual average dissolved silica concentrations in the tributaries were three- to fivefold larger that the average concentrations in the lake (Table 3-3). Annual concentrations of ammonium, phosphate, and nitrate were close to the method detection limits for those nutrients in all the tributaries (Table 3-3).

Groundwater Inflow Chemistry
Concentrations in shallow ground water (GWI) for calcium, magnesium, potassium, DIC, ANC, and dissolved silica were higher, overall, than in the lake and in the West and Northwest Tributaries (Table 3-3). The median values for GWI were smaller than the annual average lake concentrations for chloride and DOC. Median groundwater pH (6.65) was somewhat higher than the lake pH (6.55)(Table 3-3). Values for ammonium, nitrate, and phosphate in ground water were near the analytical detection limit, much like the situation for those solutes in the lake. Ground water in the watershed of Mirror Lake is the only part of the hydrologic cycle of the lake in which bicarbonate is the most abundant anion, based on equivalents (Fig. 3-3). The relative abundance of cations in ground water was similar to that in the Northwest and West Tributaries: calcium was the most abundant cation, followed in order of abundance by sodium, magnesium, and potassium.

Outlet Streamwater and Groundwater Outflow Chemistry
As anticipated, the average out-seepage groundwater (GWO) solute concentrations were similar to the lake concentrations and the outlet surface stream concentrations (Table 3-3). Small differences (<5%) existed in annual volume-weighted average values between the lake, GWO, and

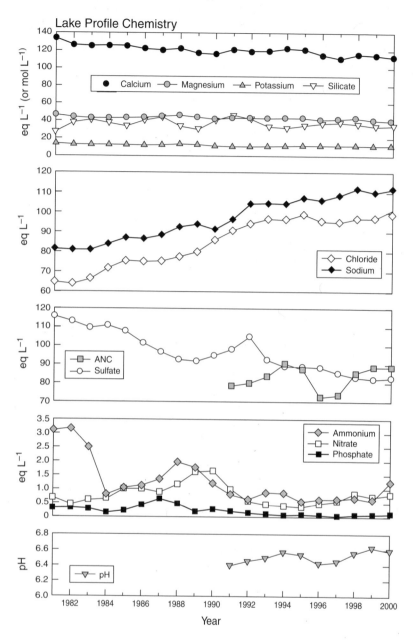

FIGURE 3-5. Annual volume-weighted average concentration time trends in Mirror Lake. Units are in µeq L^{-1}, except for dissolved silicate (µmol L^{-1}) and pH.

outlet stream values because of several factors: (1) the volume-weighted average lake concentrations included solute values from water deeper than the shallow zone of out-seepage, (2) the weekly outlet streamwater data included short-term winter ice-melt events that were more dilute and acidic than the average lake water from the same time period, and (3) monthly average outlet stream data were used for monthly GWO concentrations, resulting in a dampening of variability in GWO (see Research Methods, this chapter).

Solute Concentration Trends
Mirror Lake was the primary accounting unit for the 20-year study of hydrologic characteristics, including ground water, so we continue our description of chemical trends by first examining solute concentration changes over time within the lake itself. Measurable chemical changes in the lake provide a basis for an appraisal of input changes and a quantitative assessment of internal lake processing. Annual volume-weighted average concentrations in the lake were examined to determine if concentration distributions were representative of a normal distribution using the Shapiro-Wilk test (Table 3-4) using an α level of 0.1, which is recommended for the number of observations (years) examined (Helsel and Hirsch 1997). For those solutes with concentrations representing a normal distribution (calcium, magnesium, potassium, hydrogen ion, sulfate, DIC, ANC, and dissolved silica), statistically significant trends were based on the simple linear regression of annual volume-weighted average concentration with year. For those solutes with concentrations that did not represent a normal distribution (sodium, ammonium, nitrate, chloride, and phosphate), a Kendall tau test was done to determine if concentration and year were related.

Mirror Lake Trends
The statistical relation between concentration and year was not significant at the 99-percent confidence level (Table 3-5) for hydrogen ion, nitrate, dissolved silica, DIC, or ANC. Insufficient data were available for DOC to determine the relation to year. The declines in annual average volume-weighted concentrations of calcium, magnesium, potassium, and sulfate, as well as ammonium and phosphate (Fig. 3-5), were statistically significant.

The increases for the annual volume-weighted average concentrations of sodium and chloride (Fig. 3-5) also were statistically significant. Although the slopes of the relations were somewhat shallow, the slopes were significantly different than zero. Year accounted for more than 85 percent of the variability in concentration for sodium and chloride, and more than 70 percent of the variability in concentration for calcium and sulfate. The variability in concentration in relation to year was unremarkable for most of the other chemical constituents in lake water. The relatively short period of record for average annual concentrations of hydrogen ion, DOC, DIC, and ANC may be a factor in the absence of a detectable trend in concentration for those solutes in the lake (Fig. 3-6k). Dissolved silica data, however, do span the period that is the focus of this book (Fig. 3-6n). The declines in the annual volume-weighted averages of magnesium and potassium concentrations were not much greater than the analytical precision, and ammonium, nitrate, and phosphate concentrations were near the method detection limit for these solutes.

Bulk Precipitation Trends

Some of the solute trends observed in the lake were similar to trends observed in direct atmospheric inputs to the lake. For example, annual average hydrogen ion and sulfate concentrations declined in general in precipitation (Fig. 3-6f and g), as was the case for the lake. In precipitation, the decline was statistically significant for hydrogen and sulfate at the 95-percent confidence level, but only a third of the variability in concentration of hydrogen ion or sulfate was related to year (Table 3-5). On the other hand, the annual volume-weighted concentration of potassium and ammonium increased in precipitation (Table 3-5). For the remainder of the solutes examined in precipitation, no trend was evident in precipitation or statistically significant between 1981 and 2000.

Trends in Inflowing Tributaries

In the tributaries flowing into the lake, some annual volume-weighted concentrations did change significantly between 1981 and 2000 (Fig. 3-6a), and as noted previously, these declines or increases were consistent with the occurrence of declines or increases in the lake. For example, based on regression coefficients (r^2) or Kendall tau correlation coefficients (r) for annual average volume-weighted sodium concentrations, year was

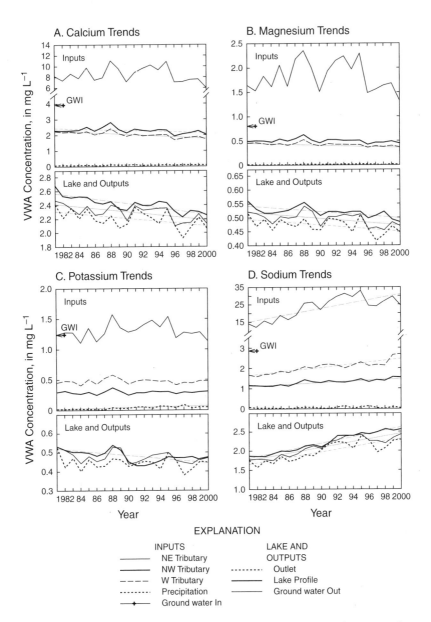

FIGURE 3-6. Annual volume-weighted average concentration (mg L^{-1}) time trends by solute (calcium to silicate) for bulk precipitation, tributary water in, ground water in, stream water out, ground water out, and lake water. Regression lines are shown for slopes that are significant at $P < 0.05$ (see Table 3-4 for complete statistics).

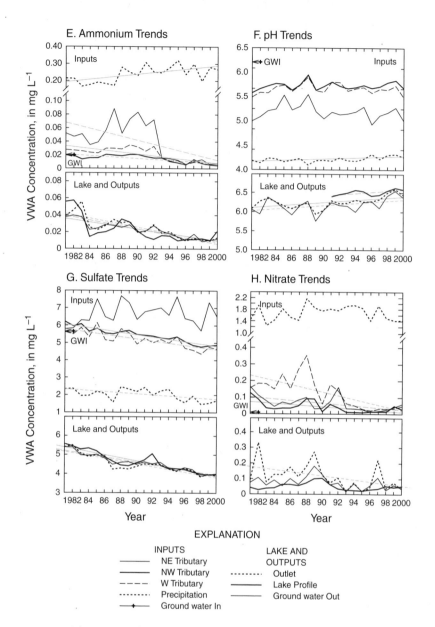

FIGURE 3-6 (CONTINUED).

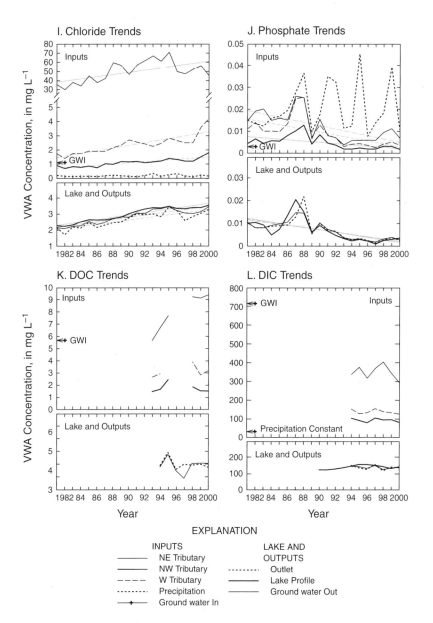

FIGURE 3-6 (CONTINUED).

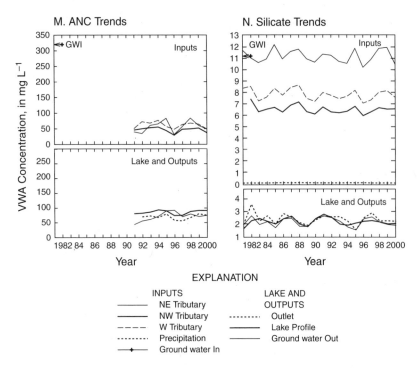

FIGURE 3-6 (CONTINUED).

statistically related to 80, 76, and 61 percent of the variability in the West, Northwest, and Northeast Tributaries, respectively. For chloride concentration, year was related to 74, 79, and 36 percent of the variability in the West, Northwest, and Northeast Tributaries, respectively. For the noteworthy declines in calcium concentration in the tributaries, year was related to 23 percent of the variability in the West Tributary, was related to 66 percent of the variability in the Northwest Tributary, and was unrelated in the Northeast Tributary. For sulfate concentration, which declined in the lake, variability in year was associated with 61 percent and 91 percent of the variability in the West and Northwest Tributaries, repectively, and not at all in the Northeast Tributary.

Annual volume-weighted concentrations of ammonium, nitrate, and phosphate all were present at concentrations in 2000 that were somewhat smaller than those present in 1981 (Fig. 3-6e, h, and j). The change in ammonium concentration was most pronounced in the Northeast

Tributary when comparing 1981 to 1992 and 1993 to 2000. A similar pronounced change was observed in the Northeast Tributary for nitrate data as well, with larger annual volume-weighted concentrations prior to 1992 and smaller values from 1993 forward. Phosphate concentrations were generally smaller in tributaries after 1992 than prior years to 1981. The statistically significant declines in ammonium concentrations in the lake (Table 3-5) were similar to declines observed in the tributaries, For phosphate, however, the decline in the tributaries was not well defined by year, as variability in year was only associated with 50 percent or less of variability in phosphate. Significant declining trends in streamwater nitrate were not matched by significant trends in lake nitrate concentrations. Much of the observed decreases in streamwater ammonium and phosphate concentrations, however, occurred after changes in analytical equipment (see Research Methods, this chapter), which suggests cautious interpretation of these results (see Discussion Question 3).

Groundwater Inflow Trends
Unlike other components of the hydrologic cycle of Mirror Lake, groundwater data were not collected on a regular basis at uniform intervals. Instead, water was collected from most of the shallow observation wells present in the watershed, providing useful information on the spatial variation in chemical characteristics of shallow ground water (Fig. 3-7). Spatial variation in chemical characteristics among wells was greater than temporal variation in chemical characteristics of the few wells from which more than one sample was collected. Some of the spatial variation in chemical characteristics of shallow ground water in the Mirror Lake watershed is related to the position of wells in the flow system. Wells in areas that recharge ground water contain lower concentrations than wells in areas where ground water has been in contact with the till for some time.

Trends in Outlet Surface Water and Groundwater Outflow
As expected, trends in the annual volume-weighted average concentration in the Mirror Lake surface water outlet and groundwater outlet reflected trends observed in the lake with few exceptions (Tables 3-5 and 3-6). Calcium and magnesium concentrations declined slightly, though significantly, in both outflows, whereas sodium concentrations increased markedly. For major anions, sulfate concentrations declined and chloride

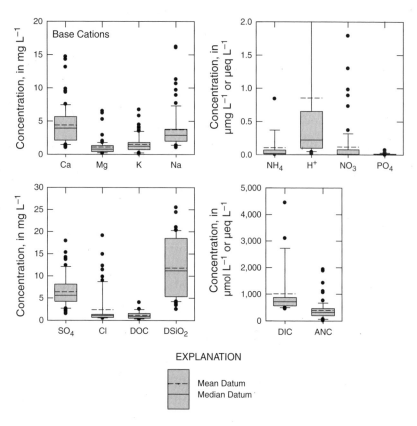

FIGURE 3-7. Box plots used to show the median value of all ground water.

increased at the outlet and in the GWO. Values of pH also generally increased between 1981 and 2000. Ammonium and phosphate concentrations declined as well. No obvious trend was evident for potassium or dissolved silica concentrations in either outflow.

Nitrate also was remarkable because, unlike most other chemical constituents, the annual volume-weighted average concentration in the surface water outlet often was noticeably greater than that found in the lake or in groundwater outflow, and the 20-year decline in surface outlet nitrate was significant, though highly variable (Fig. 3-6h). In some years, the weekly outlet sampling captured the short meltwater events, which had high nitrate concentrations and high flows. The sampling for the lake chemical profile is infrequent, so this nitrate snowmelt event is typically not captured. The larger annual nitrate in the surface water outlet was a factor of short-term ice melt–water events. These events coincided with brief spikes in water discharge, resulting in daily volume-weighted values skewed strongly toward higher nitrate concentrations. In the determination of annual groundwater outflow concentrations, such short-term chemical excursions were dampened by the use of average *monthly* surface water outlet concentrations driven by relatively uniform *monthly* groundwater outflow.

Summary of Concentration Trends

Trends in concentration are a useful starting point to begin to understand the observed trends in concentration in Mirror Lake. Trends in concentration in a single solute source that were consistent with the concentration trends observed in the lake, however, may have represented only a small fraction of the total input of a solute to the lake or may only have had a minor affect on the concentration of the lake compared to concentration changes due to the biogeochemical dynamics within the lake. Thus, it was critical to examine the combination of concentration multiplied by water volume (chemical mass) to understand patterns and processes for the solutes found in Mirror Lake.

SOLUTE MASS AND FLUXES

Solute Mass in the Lake

Owing to the fact that lake volume did not change significantly in the period 1981 to 2000, the long-term trends previously described for the annual volume-weighted average concentration data in the lake profile usually resembled the total solute mass values in the lake. Calcium, magnesium, potassium, ammonium, hydrogen ion, sulfate, and phosphate mass all declined during the 20 years (Fig. 3-8a–n). For these same solutes, and using 1981 and 2000 as end points in the trends, the changes represent declines of 20, 16, 28, 145, 67, 41, and 125 percent, respectively, of the mass in the lake from the beginning to the end of the study. The mass of sodium and chloride in the lake increased substantially over time, representing a change of 38 percent and 59 percent for each solute, respectively. There was no evidence for distinguishable trends in chemical mass in the lake for DIC and ANC or for nitrate or dissolved silica between 1981 and 2000. Routine measurements of DOC from the lake profile were not made before 2000, but DOC measurements from the outlet surface flow

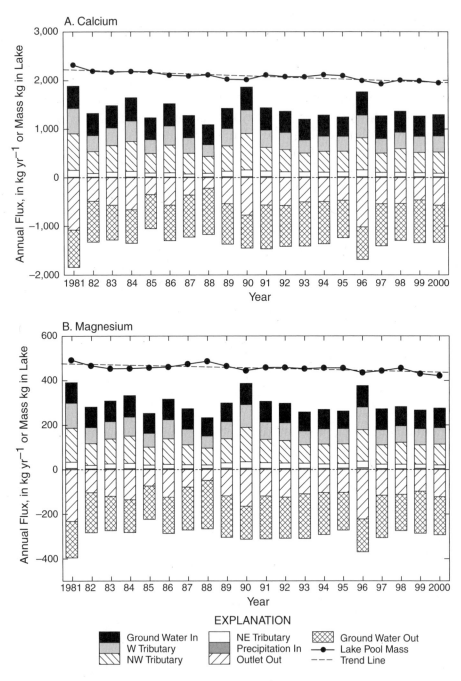

FIGURE 3-8. Time trends in annual input-output balances and lake storage changes for solutes.

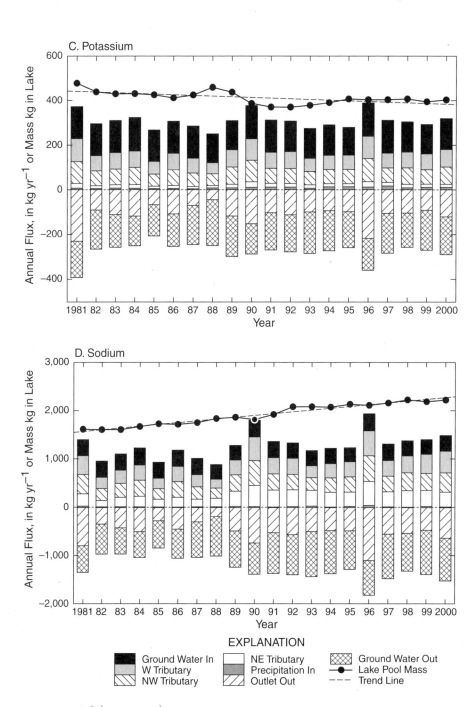

FIGURE 3-8 (CONTINUED).

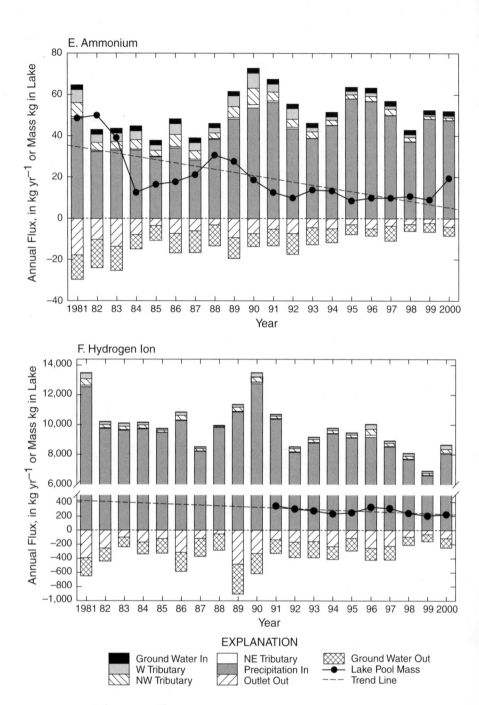

FIGURE 3-8 (CONTINUED).

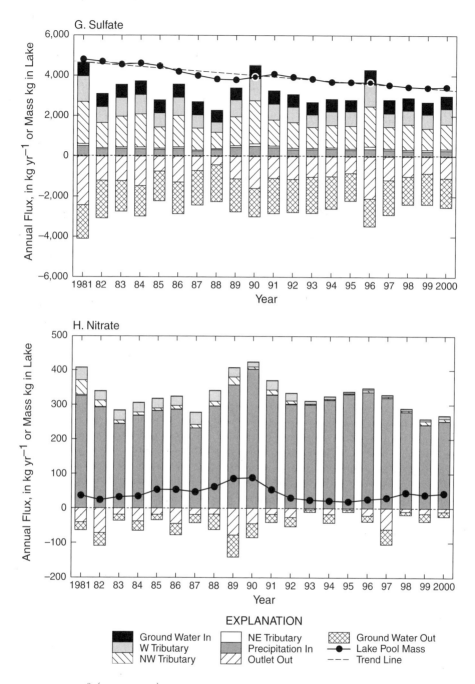

FIGURE 3-8 (CONTINUED).

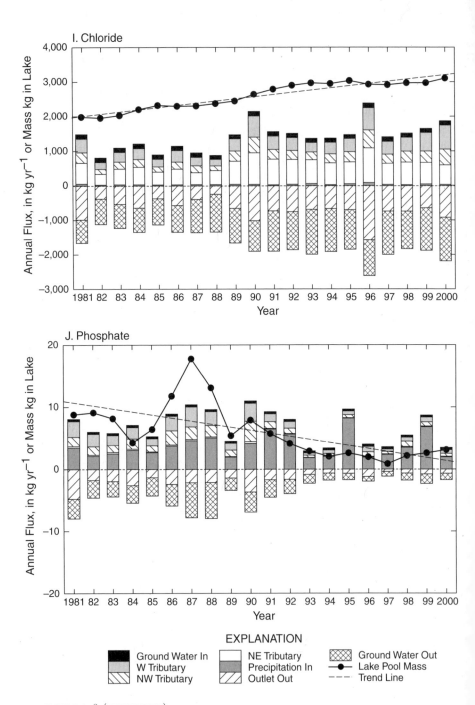

FIGURE 3-8 (CONTINUED).

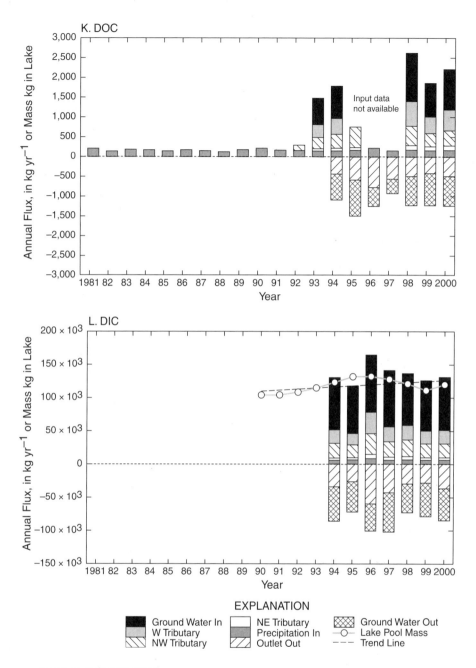

FIGURE 3-8 (CONTINUED).

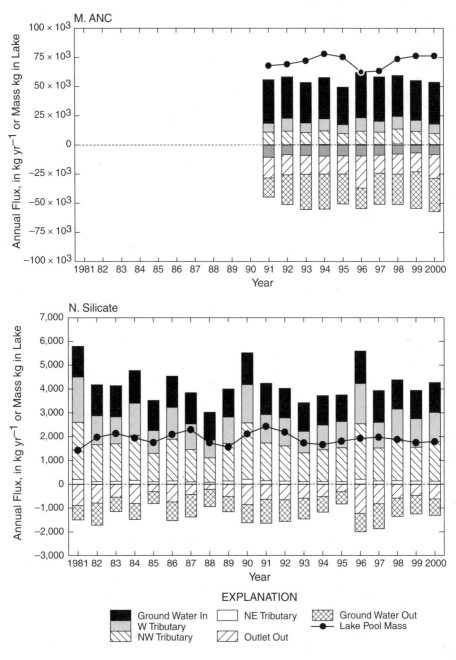

FIGURE 3-8 (CONTINUED).

from the lake (averaging 2.2 ± 0.6 mg L^{-1}), excluding winter large-flow events (avoiding surface meltwater episodes), indicate that the lake DOC mass was probably steady from at least 1994 through 2000.

Precipitation Influxes
Direct precipitation to the lake was 26 percent of the total annual inflow, but contributed most of the annual mass flux of nitrate (91% of total average inputs), ammonium (82%), and phosphate (55%)(Table 3-7a). The large contribution of precipitation to these annual solute fluxes reflects the fact that concentrations of those solutes were much larger in precipitation than in the tributaries or ground water (Fig. 3-6e, h, and j). For calcium, magnesium, sodium, and dissolved silica, which were present only in small concentrations, the annual precipitation influxes were less than 2 percent of the total inputs of each.

Streamwater Influxes
The three tributaries combined made up 59 percent of the total annual average water influx between 1981 and 2000 (chapter 2). This tributary water input composed the largest fraction of the annual mass flux into the lake, representing 68 percent of the calcium, 68 percent of the magnesium, 52 percent of the potassium, and 73 percent of the sodium annual average mass input (Table 3-7; Fig. 3-8a–n). Tributary water also contributed 69, 88, 78, and 70 percent of the annual average sulfate, chloride, DOC, and dissolved silica inputs, respectively. Tributary water was a modest source of ammonium (14% of total annual mass flux) and nitrate (9%) to the lake, whereas the tributary contribution to annual hydrogen ion was virtually nil (<0.5%) As noted previously, the three tributaries each had unique chemical characteristics that changed over time (Fig. 3-4a, b, and c). In the 20 years of study, the contribution of the Northeast Tributary to water input to the lake was consistently less than 2 percent: ranging from 1.3 percent in 1987 to 1.9 percent in 1990, 1996, and 1997. Between 1981 and 2000, the Northeast Tributary contributed between 13.8 percent (1982) and 27.7 percent (1993) of the sodium input and 36.0 percent (1987) and 50.6 percent (1993) of the chloride input.

Groundwater Influxes
Groundwater inputs, which were only 16 percent of total annual inflow, delivered the largest single share of ANC (63% of total inputs) and DIC

(58%), and a substantial portion of calcium (31%), magnesium (30%), potassium (45%), sodium (25%), and sulfate (20%) input each year (Table 3-7; Fig. 3-8a–n). Average, volume-weighted concentrations of ANC and DIC in ground water in the lake's watershed were more than three times larger than in other budget components (Fig. 3-8l and m), enhancing the modest annual groundwater influx value. For influxes of ammonium, chloride, phosphate, and DOC, the groundwater influx was less than 10 percent of the annual mass flux, and the contribution to total annual hydrogen ion and nitrate mass flux to the lake by groundwater influx was practically zero (<0.5%).

Solute Outputs from the Lake

The majority of annual average solute export from Mirror Lake occurred as groundwater out-seepage. The proportions of annual surface outlet and groundwater outflow fluxes were relatively constant across all the solutes (Table 3-7; Fig. 3-8a–n). From 51 to 59 percent of the average annual lake exports for calcium, magnesium, potassium, sodium, ammonium, hydrogen ion, sulfate, chloride, phosphate, DOC, DIC, ANC, and dissolved silica were through groundwater outflow, whereas 41 to 49 percent were via the surface outlet. The only exception was for nitrate, where the relation reversed to 48 percent for groundwater outflow and 52 percent from the surface outlet. This small reversal in the proportion of annual exports was due to the influence of brief, high-flow, large nitrate concentration events in the surface outlet surface waters during the winter.

Relation between Water Flux, Concentration, and Solute Flux

In general, higher inputs of water to the lake resulted in higher influxes of solutes (Table 3-8a–n). For most of the solutes, the correlation between total water in and total mass flux to the lake was highly significant statistically, based on the Kendall tau correlation coefficient ($r > 0.74$, $P > F < 0.001$). The Kendall tau test was used because annual total inflow was not representative of a normal distribution. However, in cases where a single hydrologic source was the primary driver for an influx to the lake, the overall relationship between chemical and hydrologic influx became much less predictable. For example, because direct precipitation was the primary source of ammonium, hydrogen ion, nitrate, and phosphate, the correlations with total inflow (including tributary water and groundwater influx) were much weaker ($r < 0.33$) and were not statistically significant at the 95-percent confidence level, except for ammonium. A major single source for DIC and ANC to the lake was groundwater input, and because groundwater inflow was nearly constant, these solutes were not significantly related to total inflow. Surprisingly, although sodium and chloride were significantly correlated with total inflow at the 95-percent confidence level, the correlations were only 0.55 and 0.45, partly because one of the major sources of sodium and chloride mass flux (22% and 42%, respectively) was the Northeast Tributary, which only contributed 2 percent to the total water input, and because sodium-chloride concentration in direct precipitation is very, very small (Table 3-7; Fig. 3-8d and i). Furthermore, there were two- to threefold increases in sodium-chloride concentrations in all three tributaries over 20 years, which tended to complicate long-term correlations between water flux and solute flux.

On the outflowing side of the lake, the surface outlet and groundwater outflow chemical fluxes combined were correlated with annual water outflux for calcium ($r = 0.72$), magnesium ($r = 0.76$), potassium ($r = 0.71$), hydrogen ion ($r = 0.33$), and sulfate ($r = 0.48$), based on the Kendall tau correlation coefficient at the 95-percent confidence level. Total annual water outflow did represent a normal distribution, but chemical fluxes for these solutes did not, as was the case for phosphate, which was not significantly correlated with outflow. Exports for solutes representative of a normal distribution, with a very strong seasonal affinity for uptake or retention in the lake, such as ammonium, nitrate, phosphate, DIC, and ANC, were not significantly related to total outflow of water based on Pearson correlation coefficients. Dissolved silica export, also normally distributed, was significantly correlated (Pearson $r = 0.74$) at the 95-percent confidence level. Dissolved silica concentrations declined at the largest outflows, as there is a strong negative relationship between tributary water dissolved silica and streamflow. Export of sodium (Pearson $r = 0.79$) and chloride (Pearson $r = 0.69$) were significantly correlated to total outflow. The large, gradual increases in sodium and chloride concentrations, and the steady decrease in sulfate concentration in the lake over time, obscured some of the comparatively modest variability in annual outflow. In these latter three cases, the short-term concentration changes during different flow conditions were smaller than concentration changes observed between 1981 and 2000.

Chemical Budgets

Chemical budgets of a lake involve accounting for inputs, outputs, and change in mass within the lake. The change in mass within the lake, also referred to as change in storage, results from the difference between hydrologic inputs and outputs, as well as non-hydrologic sinks and net gaseous flux. A complete chemical budget, or mass balance, also is termed net ecosystem flux (Likens 2002; chapter 1). Comparison of chemical mass input and output associated with hydrologic processes sets the stage for developing hypotheses concerning changes that might be expected to occur in the storage or mass within the lake based on the comparison. Such hypotheses are tested by comparing the difference between mass input and output to changes measured independently in chemical mass within the lake, storage. For the purposes of our discussion of the chemical budgets of Mirror Lake, the comparison of chemical mass input and output associated with hydrologic processes is referred to as the input-output balance. Of course, inputs rarely will be equal to outputs (the case in which the input-output balance is zero). When the annual input-output balance is positive, meaning more chemical mass enters the lake than leaves the lake by hydrologic fluxes, our expectation is that the mass within the lake should increase. When the annual input-output balance is negative, meaning more chemical mass leaves the lake than enters the lake by hydrologic fluxes, our expectation is that the mass within the lake should decrease.

The discussion that follows is an analysis of the results found when annual inputs and outputs at Mirror Lake were used to determine annual input-output balance, and then compared to independently determined changes in chemical mass in the lake itself. Change in chemical mass in the lake was a function of chemical inputs, losses, and processes within the lake. The calculation of input-output balance provided a useful framework for the evaluation of measured changes in chemical mass in the lake. If the change in solute mass was greater than expected based on the difference between external inputs and outputs from the lake, then a source of the solute may have been unaccounted for, or the solute may have been generated by internal lake processes. If change in lake solute mass was below that expected based on the difference between external inputs and outputs from the lake, then a loss term may have been unaccounted for, or the solute may have been removed from the water column by an internal process, such as sedimentation or gas efflux.

We recognized that uncertainty also needs to be factored into the complete mass balance budget of the lake. Obviously, uncertainties existed in the determination of solute inputs, losses, and change within the water column of the lake, and unmeasured fluxes, such as gaseous losses or sedimentation, were a part of that uncertainty term. These reactions made the solute budgets somewhat more complicated than the water budget. Thus, the remainder, or residual, of the solute budget was more accurately expressed as

$$\text{Inputs} - \text{Outputs} - \text{Change in Mass}$$
$$= \text{Uncertainty and Unaccounted Processes} = \text{Residual}$$

The resulting residual term (residual) often was a value other than zero (Table 3-8, last column). In fact, the residual value tended to fluctuate between positive and negative values on an annual basis. A more complicated form of the previous equation (using the fluxes already discussed previously) can be expressed as

$$P + GWI + SWI - SWO - GWO \pm \Delta S$$
$$= [\text{Unmeasured Flux} \pm \text{Uncertainty}]$$

where

P is direct precipitation,

GWI is groundwater inputs,

SWI is surface water inputs,

SWO is surface water outputs,

GWO is groundwater outputs, and

$\pm \Delta S$ is change in mass in lake.

Clearly, a full discussion of the complete chemical budget for any solute required examination of the effect of uncertainties in each parameter on the interpretation of the data. Rather than elaborate on that complex issue here, uncertainty is addressed as a separate subject in chapter 5. However, it should be noted that with the emphasis on improved determination of groundwater contribution to the complete hydrological budget of the lake between 1981 and 2000, the solute budgets were subject to less uncertainty than for budgets determined prior to 1981 (Likens 1985).

The Relation of Inputs to Outputs

Annual inputs to and outputs from the lake, averaged over the 20 years of 1981 to 2000, are shown in Fig. 3-9. Precipitation contributed the single largest fraction of the solute balance of the lake for hydrogen ion (91%), nitrate (78%), ammonium (64%), and phosphate (35%) despite the fact precipitation contributed, on average, 13 percent of the water budget (Table 3-9). The large contribution of precipitation to those solute balances reflects the fact that concentrations of those solutes were much larger in precipitation than in the tributaries or ground water (Table 3-3). The tributaries flowing into Mirror Lake contributed the largest fraction of the solute balance for silicate (52%), DOC (40%), chloride (39%), sulfate (36%), sodium (36%), calcium (34%), and magnesium (33%). Groundwater flow into Mirror Lake, which was 8 percent of the water budget, contributed the largest fraction of the solute balance for ANC (34%) and DIC (36%). Average volume-weighted concentrations of ANC and DIC in ground water were more than three times larger than in other balance components (Table 3-3). Note that ANC was determined by titration, and DIC by gas chromatography (Buso et al. 2000); thus, these characteristics were independently determined. Potassium was the only solute for which a loss term accounted for the single largest fraction of a solute balance, with approximately 28 percent accounted for by lake flow to ground water.

Those solutes for which precipitation represented the greatest input to Mirror Lake—hydrogen ion, nitrate, ammonium, and phosphate—were consistent in that measured inputs were greater than measured outputs (Table 3-8; Fig. 3-9). The large imbalance between input and output of dissolved silica was due primarily to input to the lake from the tributaries, accounting for more than 50 percent of the total solute budget. Groundwater input was the largest budget component contributing to marked solute imbalances resulting from more input than output for both ANC and DIC. The inputs for sulfate and potassium were larger than outputs, but the imbalance was less pronounced than for other solutes. Average values for solute fluxes indicated that inputs and outputs were nearly balanced for some solutes in Mirror Lake, including calcium, magnesium, and sodium, with the closest balance occurring for magnesium. Chloride was the only solute, on average, that had more of the solute leaving the lake than entering in a year between 1981 and 2000.

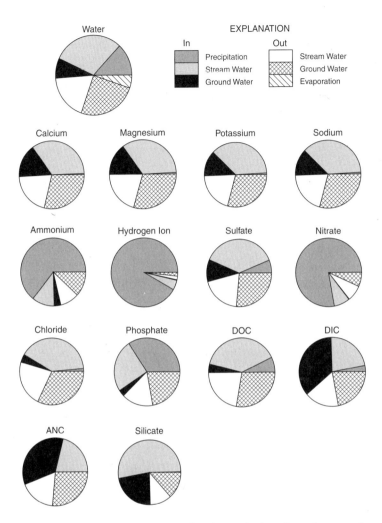

FIGURE 3-9. Average proportion of total annual solute fluxes (inputs and outputs together) contributed by each measured vector. The water budget is also shown for reference.

Chemical Retention in the Lake

The ratio of external solute outputs and solute inputs for the lake was used to examine retention or loss of solutes on an annual basis (Table 3-8). Retention, as used in this book, is the fraction of input retained in the lake, as described in Dillon and Rigler (1974):

$$R_i = (I_i - O_i) / I_i = 1 - (O_i / I_i)$$

where

R_i is retention of solute i,
I_i is mass input of solute i, and
O_i is mass output of solute i.

Retention can be expressed as a percentage of input. When solute outputs equal solute inputs, retention is zero. In the case where solute inputs exceed solute outputs, retention is positive, and the lake gains solute mass. When solute outputs exceed inputs, retention is negative, and the lake loses solute mass.

Chemical retention data (Fig. 3-10) indicate that the lake retained phosphorus, nitrogen, dissolved silica, and hydrogen ion input in each of the 20 years of study. Except for phosphorus, the percent retention was commonly more than 50 percent of input to the lake for these substances. In contrast, annual retention of calcium, magnesium, sodium, potassium, sulfate, and chloride were consistently less than 40 percent of input to the lake. In some years, this same group of chemicals also had negative values of retention. In those years, the lake lost more calcium, magnesium, sodium, potassium, sulfate, and chloride than it received in a year. More chloride left the lake by ground water and the surface water outlet than was received by the lake in 19 of the 20 years of study.

The pattern in retention for calcium, magnesium, sodium, potassium, and sulfate is primarily related to water movement through the lake. In years where water inputs exceed water losses (1981, 1987, 1988, 1992, 1993, 1994, 1997, and 1999), retention values for these constituents generally approached zero or were negative. A statistically significant correlation (probability [P] of a greater absolute value of the Pearson's correlation coefficient [r] due to chance was less than 0.001) existed between the water input-output balance and the retention of calcium ($r = 0.87$), magnesium ($r = 0.88$), sodium ($r = 0.92$), chloride ($r = 0.88$), and sulfate ($r = 0.85$). Most of the retention for these chemical constituents is a function of the imbalance between water inputs and outputs. In years when more water enters the lake than leaves, the lake retains calcium, magnesium, sodium, chloride, and sulfate. In years when more water leaves than enters the lake, the lake exports calcium, magnesium, sodium, chloride, and sulfate.

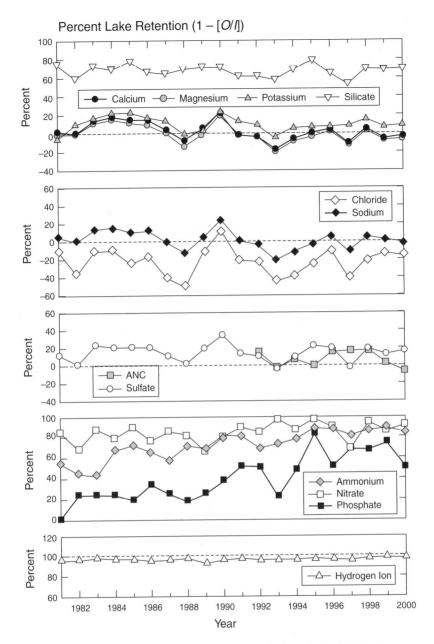

FIGURE 3-10. Annual trends in percent retention of solutes in the lake based on mass fluxes and the formula 1−[O/I].

Change in Mass and Chemical Retention

As noted in chapter 2 on the water balance of the lake, the water volume obtained by subtracting the annual volume of water output from the annual volume of water input is not the same as the independently measured change in volume. Independently measured annual change in lake volume was less than 1 percent of lake volume (8620 m^3) in all 20 years of the study. The water volume equal to water inputs minus water outputs was less than 5 percent of lake volume (43,100 m^3) in 8 of the 20 years of the study. The difference between water inputs and outputs was more than 5 percent in 12 of the 20 years of study. The closest agreement between the imbalance in inputs and outputs and independently measured change in lake volume was in 1982, 1987, 1989, 1991, and 1992 (Table 3-9).

The chemical mass obtained by subtracting the annual chemical mass output from the annual mass input differed from the independently measured change in mass in the lake for all of the chemical constituents examined in the Mirror Lake study (Table 3-10). Owing to the uncertainties in determining the volumes of water input and output, as well as lake volume, it would be remarkable if these values were identical. Furthermore, there is only an expectation that these values will match for non-reactive (conservative) solutes.

Uncertainty in measurement may account for the case when the independently measured change in chemical mass in the lake is not equal to the difference between annual chemical mass inputs and outputs (I–O). The likely uncertainty of the volume of the lake being ±5 percent or ±10 percent (chapter 5) may be used to approximate the uncertainty in the change in chemical mass in the lake. Thus, 5 to 10 percent of annual average mass in the lake may serve as a benchmark against which to judge whether or not values of change in mass in the lake and the difference between chemical inputs and outputs are distinguishable. In the case when independently measured change in chemical mass in the lake and the chemical input-output balance both lie within the range of ±5 percent or ±10 percent of the annual average mass, the discrepancy between the masses simply may be due to uncertainty in measurement.

Annual values of independently measured change in mass and change in mass expected based on the difference between chemical inputs and outputs (the input-output balance) indicate that calcium, magnesium, sodium, potassium, chloride, sulfate, and ANC generally were within the range of values defined by ±10 percent of the annual average mass in the lake. A noticeable disparity existed between the expected change in mass based on the difference between inputs and outputs and determined from lake concentrations and lake volume for hydrogen ion, phosphate, nitrate, ammonium, and dissolved silica (Table 3-10). The disparity is consistent with the fact the lake retains these constituents.

The annual change in chemical mass in the lake for all solutes was usually small (<4%). Yet the 20-year average input-output balance mass and 20-year average change in mass in the lake were statistically indistinguishable for calcium, magnesium, and sodium (based on a t-test comparing the 20-year average values, probability of a greater absolute value of $t = 0.01$). Biological uptake in the lake of these nutrients was likely minimal. Solutes for which the annual differences between inputs and outputs were significantly larger than the average change in chemical mass in the lake were ammonium, hydrogen ion, nitrate, ANC, and dissolved silica. The lake acted as a strong sink for these chemicals, resulting in large ratios (greater than 10:1) for input-output balance compared to change in lake storage. High year-to-year variability in lake storage of potassium, sulfate, chloride, phosphate, and DIC meant that while the average mass storage changes were typically less than the average I-O values, the differences between these values were not significant (t>0.05). The large variability in annual chemical storage may have been the result of differences in internal lake metabolic processes between years.

Chemical Input-Output Balance and Chemical Retention in the Watersheds

The watersheds of Mirror Lake retain the majority of the input of hydrogen ion (>90%), phosphate (>50%), nitrate (>90%), and ammonium (80%) to the watersheds from precipitation. The watersheds export calcium, magnesium, sodium, potassium, chloride, and dissolved silica to the lake. Aside from chloride, the export represents weathering (that is, physical and chemical breakdown of primary and secondary minerals) and leaching from the soil exchange system within the watersheds, with considerably more apparent weathering and/or exchange taking place in the Northeast Tributary, most noticeably for calcium, magnesium, sodium, and potassium. The greater apparent weathering in the Northeast Tributary is confounded by sodium-chloride from road salt that enters the watershed. The fact that the dissolved silica export (represented by negative values of retention) for each of the three watersheds is similar suggests that general weathering within the watersheds should be approximately the same.

Weathering within the watersheds also provides input of chemicals into the lake from ground water. Calculation of chemical retention within ground water is somewhat more problematic, as the exact chemical input at the water table is not known. The magnitude of the chemical input from ground water to the lake generally is similar to input from the Northwest and West Tributary watersheds to the lake (Table 3-5).

Because of nearly complete neutralization of atmospheric deposition within the terrestrial watershed, 95 percent of the annual hydrogen ion mass flux to the lake was from direct precipitation to the lake's surface. The terrestrial watershed for the lake retains the majority of ammonium, nitrate, phosphate, and hydrogen ion received from atmospheric deposition, as is the case in the experimental watersheds of the Hubbard Brook Valley (Likens and Bormann 1995).

CLIMATE INFLUENCES ON CHEMICAL SEASONAL PATTERNS

As a brief extension to the above, it is important to mention that although the annual flux data may have displayed relatively little variance over 20 years, the data derived from our monthly input and output measurements showed robust and repeatable patterns in solute flux that were very closely related to differences in seasonal climate within each year. Furthermore, mass and concentrations in the lake itself varied by season for several biologically important solutes. Thus, we finish this chapter with an overview of seasonal aspects of the 20-year record.

Change in the hydrologic regimes of some lakes in response to dry conditions, for example, can result in changes in solute input, thereby causing a marked effect on a lake's solute concentrations and seasonal solute dynamics (Krabenhoft and Webster 1995). However, the period covered by this study of Mirror Lake was considered "wet" by long-term standards, and many of the years were very similar in total amounts of precipitation (chapter 2). Dry spring seasons were often balanced by wet autumns, and such hydrologic variety was hidden in the annual summations. The typical annual hydrologic pattern was one of high total inputs and outputs in March and April (driven by snowmelt), lower inputs from

July through September, and then increased inputs and outputs in October and November. In fact, the wettest (1996) and driest (1988) years had similar and typical monthly input and output water patterns. Years with unusual maximum monthly water peaks were 1981 (peak in February), 1984 (in May), 1990 (in August), and 1998 (in June).

As expected, for many of the solutes considered, monthly fluxes paralleled monthly water flow. For calcium, magnesium, potassium, sodium, sulfate, chloride, and dissolved silicate, monthly peaks for inputs and outputs occurred in March and April, when water fluxes were maximized by snowmelt. These same solute flux patterns were repeated in both wet and dry years. The solute inputs dominated by direct precipitation—that is, ammonium, hydrogen ion, nitrate, and phosphate fluxes—tended to peak in those months when precipitation was at a maximum, and streamflow at a minimum. The most extreme example of this "pulse" effect was in 1990, when the maximum annual water input occurred in August, after a relatively dry June and July. In this case, direct inputs to the lake of ammonium, nitrate, and phosphate from this one month pushed the total inputs for the year to the highest recorded for the 20-year period. For hydrogen ion, 1990 was the second-highest year for total inputs, after 1996 (the wettest year). Extreme climatic events of this kind (a mid-summer extra-tropical storm) are a relatively rare occurrence in non-coastal New England. This demonstrates the need for long-term (decadal) monitoring in order to observe and evaluate at least a few extraordinary situations.

Within the lake itself, in the years prior to 1981, the solutes nitrate and silicate had the most noticeable monthly fluctuations in volume-weighted concentrations and mass (Likens et al. 1985). A similar observation was made during the period 1981 to 2000. The larger values of nitrate commonly occur in March; minimal values are characteristic of June to November (Fig. 3-11). Concentrations near the detection limit during many months of the year are consistent with use of this solute by biota in the lake. Use of nitrate by biota related to the minimal values between June and November further corroborate the large amount of annual input that does not leave the lake by the hydrological pathways of water loss to ground water or the surface outlet of the lake (Fig. 3-10). Thus, most of the retention of nitrate likely takes place from June to November. Furthermore, for most months, solute inputs of nitrate exceed solute losses by hydrologic outlets, to ground water or the surface outlet of the lake (Fig. 3-12). In the absence of loss (uptake) of nitrate by biological activity, the monthly imbalance would be expected to result in a steady increase in nitrate mass in the lake. Instead, the seasonal pattern in the nitrate mass in the lake is the same as for the volume-weighted concentration of nitrate (Fig. 3-13).

Dissolved silica concentrations also generally were larger in March (Fig. 3-14), like nitrate. Unlike nitrate, concentrations of dissolved silica declined gradually through August, followed by a rise in the last two months of the year. The seasonal pattern in the difference between dissolved silica inputs and outputs is similar to that of concentration (Fig. 3-15). Thus, the seasonal fluctuation in volume-weighted average concentrations of dissolved silica in the lake might appear to be primarily a function of the seasonal hydrograph of the lake, its tributaries, and outlets. When the mass input-output data are placed on the same scale, however, it becomes clear that the observed month-to-month variation in silicate mass in the lake is larger than solely attributable to the solute mass difference between hydrologic inputs and outputs alone (Fig. 3-16). The decline in silicate lake mass from May to September that is larger

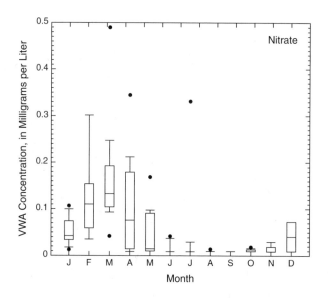

FIGURE 3-11. Monthly volume-weighted average nitrate concentrations in Mirror Lake, 1981–2000.

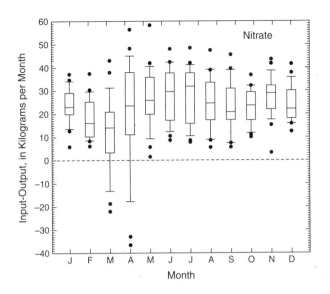

FIGURE 3-12. Monthly difference between nitrate mass input and nitrate mass loss from Mirror Lake, 1981–2000.

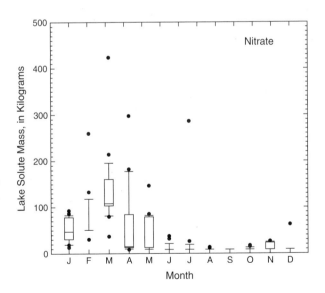

FIGURE 3-13. Monthly nitrate mass in Mirror Lake, 1981–2000.

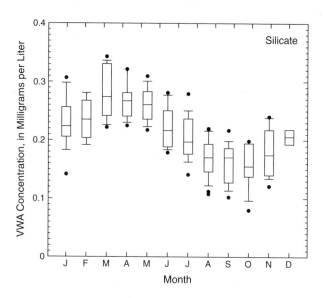

FIGURE 3-14. Monthly volume-weighted average (VWA) silicate concentrations in Mirror Lake, 1981–2000.

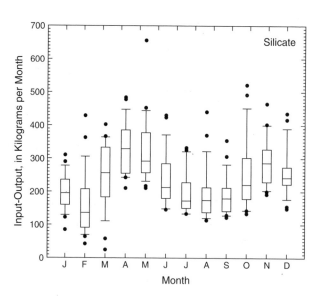

FIGURE 3-15. Monthly difference between silicate mass input and silicate mass loss from Mirror Lake, 1981–2000.

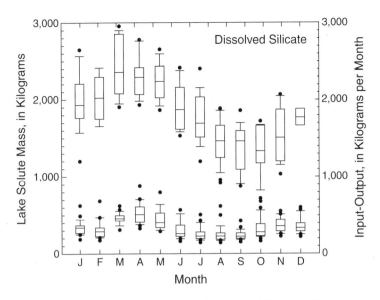

FIGURE 3-16. Comparison of monthly lake mass of dissolved silicate and the difference between silicate input and loss associated with hydrological processes.

than expected based on solute mass imbalance likely reflects biological uptake that accounts for lake retention of more than 50 percent of input on an annual basis (Fig. 3-10). Note that in Fig. 3-10 we compare the difference between solute mass input and output to lake solute mass observed in a month, not the change in lake solute mass each month.

Seasonal changes in other solutes, calcium, magnesium, potassium, chloride, sulfate, ammonium, and phosphate, are not as pronounced as for nitrate and dissolved silicate (Fig. 3-17) based on examination of the entire range of concentrations observed each month. Thus, seasonal changes in these solutes are similar to the observations made concerning seasonal changes prior to 1981 (Likens 1985). Median concentrations for each month recorded from 1981 to 2000 indicate that during the time of maximum water input in the spring, calcium, magnesium, and potassium may decline.

The lowest median value for lake ANC occurs in May after mixing of the water column, and after the lowest ANCs occur in the tributary streams in March and April. This evident lag is likely driven by late-winter stratification. Because of temperature-induced water density

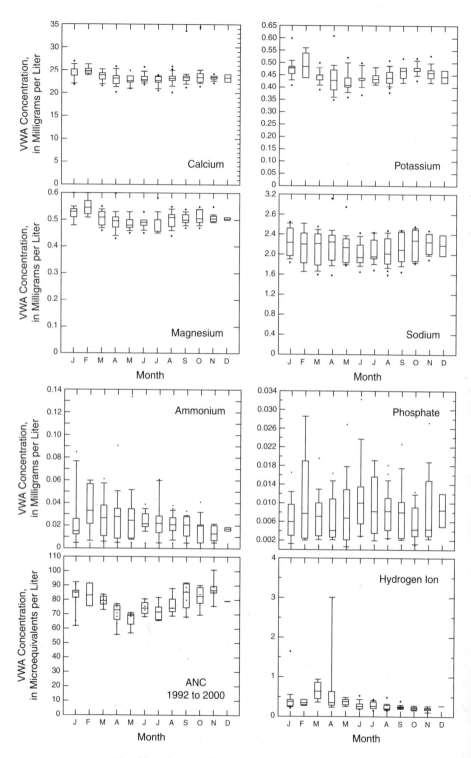

FIGURE 3-17. Monthly volume-weighted average concentrations of selected solutes.

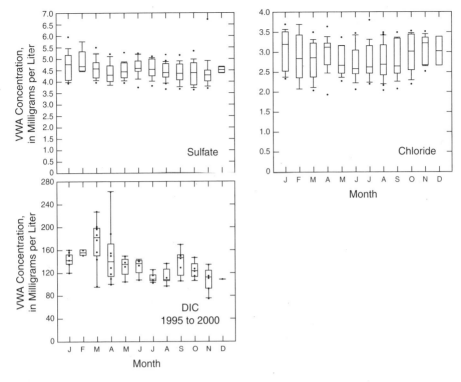

FIGURE 3-17 (CONTINUED).

differences, stratification under ice cover tends to cause shunting of the cold, dilute snowmelt water across the surface of the lake without deep mixing. The extreme values for concentration of hydrogen ion do not alter the trend evident from examination of median values for each month; values tended to be largest in March and April followed by a decline into May and little change in summer and fall.

The extreme values for each of the solutes across the months in Figs. 3-11 to 3-17 do not correspond to all of the values of either the wettest or driest years of the 20 years of interest. To illustrate this, the monthly pattern in the difference between silicate solute mass input and output can be examined (Fig. 3-18). Only in April was the largest value of input minus output associated with the wettest year (1996). Only in June and November were the smallest differences in input minus output associated with the driest year (1988). Despite the clear differences in inputs and outputs, the observed changes in net balance on a monthly basis may

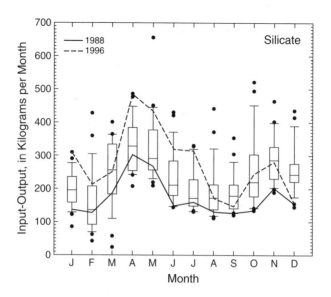

FIGURE 3-18. Monthly difference between silicate mass input and loss with dry year (1988) and wet year (1996) highlighted.

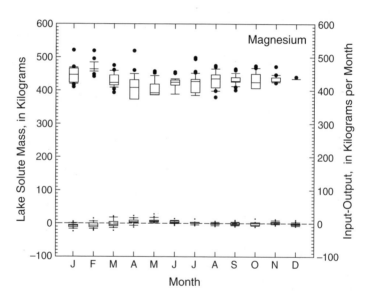

FIGURE 3-19. Comparison of monthly lake mass of magnesium and the difference between magnesium input and loss associated with hydrological processes.

only represent a fraction of the overall solute mass in the lake (Fig. 3-16). As previously noted in the case of silicate, differences in lake solute mass from month to month may be larger than expected, on the basis of a comparison of solute mass input and output associated with the movement of water into and out of the lake, owing to use of silica by biota. In the case of a solute where only a small fraction of the input may be used by biota, such as magnesium, the solute mass in the lake may not vary noticeably from month to month, and the difference between inputs and outputs is quite small in relation to the solute mass in the entire lake (Fig. 3-19).

In summary, seasonal variation in solute concentrations in the lake between 1981 and 2000 are similar to seasonal patterns reported by Likens (1985) for data collected prior to 1981. Even those solutes for which significant long-term annual trends became established since 1981 (e.g., calcium, sodium, hydrogen ion, sulfate, chloride) continued to exhibit similar seasonal fluctuations. The relatively wetter period represented by the 20 years that are the focus of this book has not changed the seasonal chemical "signature" of Mirror Lake in relation to its solutes. Within the 20-year period, wet and dry years did not define the extreme values in concentration in the lake for every month. Atypical "pulses" of inputs influenced annual values when the events were extraordinary in timing and volume, such as in August 1990, but primarily for those solutes delivered by precipitation. These data point out the importance of being aware of the seasonal and yearly hydrological context in which observed changes in solute concentrations are interpreted.

DISCUSSION

This chapter began by raising questions that will now be addressed using the 20-year data set. The primary question was "How important was it to quantify all the chemical and hydrologic fluxes in order to answer fundamental questions about Mirror Lake?" While there are no doubt many other questions the reader might suggest, we have chosen the following five as a means of demonstrating our approach and keeping this large and complex project focused.

ACIDIC DEPOSITION, RESPONSE, AND RECOVERY

How has acid deposition affected Mirror Lake, and can the response or recovery be quantified? Acid rain has been affecting the entire eastern half of the United

States and southeastern Canada, since at least the mid-1950s (Likens et al. 1970). The Mirror Lake watershed is no exception. Precipitation, tributary, lake, and even ground water at Mirror Lake are characterized by enhanced levels of sulfate (ranging from 2 to 8 mg L^{-1}). Elevated sulfur deposition originated from sulfur dioxide emissions through combustion of coal and oil, largely from electrical utilities in the midwestern United States (Likens et al. 1972; Cogbill and Likens 1974; Likens and Bormann 1974; Driscoll et al. 2001), and variation in these emissions are strongly correlated over time with precipitation and tributary water concentrations in the Hubbard Brook Valley (Likens, Buso, and Butler 2005). Background values for sulfate in precipitation and tributary water are estimated to have been no higher than 1 mg L^{-1} in New England prior to the Industrial Revolution (Likens et al. 1996). Local schist bedrock and till have a minor sulfide component (Barton et al. 1997), but where present, the surfaces of these minerals tend to be well weathered, so that geologic weathering is not likely to be a major source of sulfate currently. Furthermore, analyses of naturally occurring sulfur isotopes suggest that tributary water and groundwater sulfate are similar; they are both largely atmospherically derived and are not from local mineralogy (Likens et al. 2002).

Inputs of pollutants that can travel in the atmosphere over long distances, such as sulfate, have not only been widespread: they have been substantial. The total sulfate input to Mirror Lake for 1981–2000 was 64,686 kg, of which 7344 kg (11%) was by direct precipitation to the lake's surface. In the longer context of measurements begun in 1963 at nearby Hubbard Brook Experimental Forest, the entire 85 ha watershed (including the lake) received 94 metric tons of sulfate from bulk precipitation from 1964 to 2000. It is likely that total cumulative sulfate deposition (including dry deposition) since 1900 could have been several-fold larger (Gbondo-Tugbawa et al. 2002; Likens et al. 2002). It is not surprising, then, that sulfate represented a high proportion of the ionic charge in precipitation (27%), tributaries not heavily affected by road salt (21%–28%), and the lake (18%)(Fig. 3-2). Although the groundwater influx ionic charge distribution is dominated by ANC (36% of total charge), it also had sulfate concentrations roughly equal to that in any of the tributaries and the lake (Table 3-2; Table 3-3; Fig. 3-2). Clearly, sulfate from atmospheric deposition has permeated the entire terrestrial and aquatic ecosystem.

Nitrate, derived from emissions of nitrous oxides (NO_x) via industrial and transportation sources, is another large component of acidic precipitation (about 30%–35% of the average anion charge in bulk precipitation) (Fig. 3-2). As sulfate declined in response to U.S. Clean Air Act emission controls, the proportional contribution of nitrate increased (Likens et al. 2005). Direct precipitation to the lake was by far the dominant source of nitrate input (90.7% on average; Table 3-7a), and these fluxes were fairly uniform seasonally within a given year. Thus, the strong seasonal pattern of higher nitrate concentrations and fluxes in the spring during snowmelt, documented in the small experimental watersheds at Hubbard Brook Experimental Forest (Likens and Bormann 1995), was subdued by comparison in Mirror Lake. Moreover, with mixing and internal uptake, the lake tended to start and end each year at low levels of nitrate, such that annual lake mass changes were less than 1 percent of total inputs on average (Table 3-8h).

The average concentration of nitrate in the lake was typically less than 0.05 mg L^{-1}, which was expected since the lake is nitrogen co-limited with phosphorus (Gerhart and Likens 1975; Bade et al. 2009). Direct input of nitrate via precipitation to the lake from 1981 through 2000 was 5990 kg; total mass flux, including all inputs, was 6608 kg. Because of strong retention, presumably by biological uptake, only 1096 kg (17%) was exported from the lake in 20 years. The entire 85 ha watershed received almost 63 metric tons of nitrate via precipitation from 1964 through 2000. Biological reactivity (retention) of nitrate in the drainage basin obviously prevented serious changes in trophic conditions for the lake from this large nutrient addition (see chapter 6), because it reduced the total nitrate influx to the lake to just a fraction of that deposited on the watershed.

Partially balancing the ionic charge of sulfate and nitrate deposition, hydrogen ion inputs were also substantial at Mirror Lake. The 20-year average pH of precipitation was 4.3 during this study, and this converts to a total hydrogen ion deposition to the lake and its entire watershed of greater than 1 million equivalents. Yet acid neutralization within the watershed of the lake reduced the total hydrogen ion input to the lake to 19 percent of the deposition that entered the watershed, and most of that (95%) came in through direct precipitation to the lake's surface. Changes in lake pools of hydrogen ion were very small relative to the fluxes (<1% of influxes), and this change is evidenced by the small variability in lake pH from year to year.

How much has Mirror Lake changed due to inputs from acidic deposition? If the lake's average sulfate ionic charge (96 µeq L^{-1}) were reduced to the theoretical sulfate background level (20 µeq L^{-1}) of the past, then to match that charge balance (assuming all major ions were stable), the ANC would need to have been increased to about 160 µeq L^{-1}. Using ANC titration curves determined for Mirror Lake water samples, and based on the relationships between DIC, ANC, and pH (Wetzel and Likens 1991), at a theoretical ANC of 160 µeq L^{-1}, the average pH of Mirror Lake would have been between 7 and 7.5 prior to titration by anthropogenic atmospheric acids. The average pH of Mirror Lake is currently 6.5.

Significantly increasing trends ($P > F = 0.047$) in the lake's annual average pH (and in the longer pH data from the outlet: $P > F = 0.040$) suggest that the system is slowly recovering from the affects of acidic deposition. Two direct reasons for this recovery are declines in hydrogen ion and sulfate inputs. The annual pH of direct precipitation increased significantly ($P > F < 0.005$) by about 0.2 units. This modest change in pH resulted in a decline in annual acid influx from 12,536 eq yr^{-1} in 1981 to 8044 eq yr^{-1} in 2000. Alternatively, based on a linear regression equation, there was a significant decline ($P > F = 0.004$) in hydrogen ion concentration of 3086 µeq L^{-1} over 20 years (Table 3-4). Sulfate concentrations in direct precipitation also declined significantly (0.6 mg L^{-1}; $P > F = 0.013$)(Table 3-4). This concentration change produced a reduction in flux from 529 in 1981 to 302 kg yr^{-1} in 2000, or about 43 percent in annual flux. As calculated from the regression equation, the decline is about 120 kg or 2496 eq over 20 years (Table 3-4). The 20-year decrease in sulfate on an equivalent basis using the regression equation, was only slightly less than that for hydrogen ion (3086 eq) using the same approach, showing the linkage between the two.

Other processes may be affecting acid-base relationships in the lake and affecting recovery from acidic deposition. These are complex and difficult to quantify, but the solute mass balance approach taken here has provided us with critical information to assess certain possibilities. However, in some instances we can only speculate. The lake average pH and ANC for 1981–2000 (pH 6.5 and +80 µeq L^{-1}) are well above the typical pH (5.2) at which ANC equals 0 in these dilute waters (Buso et al. 2000), and it is among the most well-buffered sites in the Hubbard Brook Valley (Likens and Buso 2006). Obviously, the substantial inputs of ANC from ground water and the tributaries have mitigated some of the acid inputs (Table 3-6A; Fig. 3-9). Yet internal biogeochemical processes could also have influenced the balance between acids and bases, so this is not just a simple titration.

For example, while the matching time periods are shorter than the 20-year record, the hydrogen ion and ANC imbalances are very different (Table 3-7a–n; Table 3-8). The average annual difference between inputs and outputs for hydrogen ion during 1992–2000 was 8714 eq yr^{-1} (SD = ±1483 eq yr^{-1}), which suggests considerable neutralization within the lake and relatively little year-to-year variability. Conversely, the annual average difference between input and output of ANC was 3551 eq yr^{-1}, which is much smaller and positive, implying a net loss of ANC in the lake, although there is great year-to-year variability in this difference (SD = ±4703 eq yr^{-1}). This annual variability may be a function of large and variable external inputs of ANC (average = 48,157 eq yr^{-1}; SD = ±3775 eq yr^{-1}) that dominate the ANC balance such that internal processes are obscured. It might be expected that ANC and hydrogen ion would be of equal magnitude, but as the standard deviation in the input-output balances for these solutes indicates, the year-to-year variability may be a factor in the lack of exact agreement in their input-output balances.

Several biogeochemical reactions that generate or consume ANC are distinctly possible in Mirror Lake (Likens and Moeller 1985). Unfortunately, complete ANC data are only available for the period 1992 to 2000, so comparisons with other solute fluxes and imbalances are restricted to a shorter period than the full record. An example of four potential processes follows:

1. Sulfate retention averaged 317 kg yr^{-1} (for 1992–2000), which represents 6594 eq yr^{-1} of ANC. Indeed, this equals the amount of sulfate loss corroborated by sediment studies at Mirror Lake (Giblin 1990). In the anoxic hypolimnion and sediments, bacterial sulfate reduction could have provided ANC to the lake. Fe(II) was detected in the hypolimnion of Mirror Lake (~2 mg L^{-1}); therefore, the production of iron sulfides and permanent loss of sulfur to the sediments was likely.

2. Assimilatory nitrate reduction (1992–2000) could have contributed about 3461 eq yr^{-1} of ANC, based on an average nitrate retention of 270 kg yr^{-1}.

3. Ammonium uptake (40 kg yr^{-1} average for 1992–2000) would have released hydrogen ion and consumed ANC, so that the average annual ammonium retention represents a sink of 4444 eq yr^{-1} for ANC.

4. To balance the average chloride discrepancy (release) of 435 kg yr^{-1} for 1992–2000 with an unmeasured matching cation, it is postulated that about 343 kg yr^{-1} of Fe(II) would have to enter the lake (see Discussion Question 4 below). If this reduced iron were entirely oxidized in the water column, it would have removed or consumed about 12,270 eq yr^{-1} of ANC.

Manganese might have played a role similar to iron in consumption of ANC. However, we have not constructed budgets for iron or manganese and can only speculate on their roles in the ANC budget. Potentially, hydrogen ion could have also accompanied the unmeasured chloride, directly influencing ANC. Finally, dissolution of calcium or magnesium from the sediments in response to hydrogen ion input is a possibility, though the annual imbalances in these solutes do not suggest large releases of either from such reactions in the lake.

These exercises in attempting to balance the ANC inputs aptly demonstrate the possibilities that exist, through the results of our mass balance approach. This approach also helps to identify and constrain further research into quantifying and predicting the response of the lake to acidic deposition. Clearly, the lake has been affected by acidic deposition, although there is considerable acid neutralization occurring in the ecosystem. Mirror Lake is recovering from acid rain impacts, albeit slowly. Since the terrestrial watershed has been stable and efficient in neutralizing acids and exporting bases, continued recovery appears to depend rather directly on changes in precipitation brought on by further reductions in national and regional SO_2 and NO_x emissions.

HYDROLOGIC PATHWAYS AND SALT CONTAMINATION

How have various hydrologic pathways contributed to salt contamination of Mirror Lake? The major chemical composition of Mirror Lake is not the same as it was when first sampled nearly 40 years ago, primarily because a major highway was constructed through the lake's watershed and because large amounts of salt are added to this highway during winter. A section of U.S. Interstate Highway 93 (I-93) was constructed through the east side of the Mirror Lake watershed (Northeast Watershed) in 1969–1970. Road salt pollution poses a serious problem along I-93 because the application rates are among the highest in New Hampshire, ranging from 50 to 104 metric tons per year (average 78) of sodium-chloride per km of highway (all four lanes combined; New Hampshire Department of Transportation, Bureau of Environment records 2007). To avoid direct runoff of highway drainage waters to the lake through the Northeast Tributary, 88 percent of the Northeast Tributary watershed was diverted through culverts and along an earthen berm away from Mirror Lake and toward the Pemigewasset River in 1969 (Fig. 1-3). Nevertheless, within a few weeks, the chemical characteristics of the truncated Northeast Tributary had begun to reflect the effect from construction disturbance with increased calcium and magnesium concentrations. Tributary flow was likely greatly reduced also, but an automated stream gage was not installed until 1979. That portion of I-93 within the former Northeast Watershed was opened in August 1970. This opening was followed within five years by elevated sodium and chloride concentrations in the Northeast Tributary, ostensibly from highway deicers infiltrating the ground water along the highway and seeping through the berm into the remaining Northeast Tributary.

Between 1970 and 1980, Northeast Tributary sodium and chloride concentrations increased steadily from background levels (~1 mg L^{-1} and 0.5 mg L^{-1}, respectively) and equivalent ratios (Na:Cl ~3) typical of pristine conditions in the Hubbard Brook Valley (Likens and Buso 2006) to concentrations up to an order of magnitude higher and equivalent ratios resembling sea salt (Na:Cl = 0.8). Groundwater wells (water table wells, piezometers, and bedrock bore-hole wells) located in the Northeast Tributary watershed quickly acquired elevated sodium and chloride concentrations, indicating deep penetration of road salt runoff into soils, till, and geologic fractures. Although the drainage modifications in the Northeast Watershed were estimated to have diverted about 97 percent of the mass of road salt applied to I-93 away from the lake (Rosenberry et al. 1999), the concentrations of sodium and chloride in the lake tripled by 1995. Although the Northeast Tributary represented just 3 percent of the annual average streamflow into the lake, it eventually contributed about 50 percent of the annual average influx of chloride. The receipt of sodium chloride in the Northeast Tributary from road salt not only increased sodium inputs, but induced higher calcium and magnesium influx to the lake as well, apparently through cation exchange in saturated soils and the berm (Rosenberry et al. 1999), although other sources of cations are possible. The salt used by the New Hampshire Department of Transportation (NH-DOT) is imported from halite mines in Canada and from evaporative deposits in South America (NH-DOT Bureau of Materials Report 1974). The evaporate salts are about 99 percent sodium-chloride, while the mined salt can have up to 5 percent by weight of calcium, magnesium, sulfate, and iron present. Calcium-chloride is also added directly to specific sites along I-93, such as culvert drains, but in amounts too small to influence annual ion ratios (a few kg per season; NH-DOT Bureau of Materials Report 1974).

After the first few years of highway use, the Na:Cl equivalent ratio declined to 0.4, well below the sea salt ratio, as the cation charge needed to match the chloride was made up by the other solutes such as calcium and magnesium. During 1981–2000, the ratio returned gradually toward 0.8, suggesting that over time the availability of calcium or magnesium in Northeast Tributary soils had been reduced through constant exchange with sodium. Although some of these changes might have been due to variation in the chemical quality of the road salt, the changes in the proportions of the solutes were much larger than the few percent provided from trace contaminants.

Concentration and flux values of sodium and chloride for the Northeast Tributary had a complicated seasonal pattern. Concentrations always peaked in mid-summer, and displayed a parabolic relationship with flow—that is, the maximum concentrations occurred in the mid-to-low range of the flow values. At the very smallest flows, the salt concentrations decreased due to an increasing proportion of lower-concentration ground water entering the Northeast Tributary. At the highest flows, the concentrations declined because of dilution by precipitation. Mid-summer salt levels in the Northeast Tributary may have been elevated by concentration through evapotranspiration of tributary water by the streamside forest (Rosenberry et al. 1999). Because of this pattern in concentration, fluxes to the lake peaked at the highest flows in the spring, during snowmelt, and in the fall with heavy rain from extra-tropical storms, but not during the winter when the salt was applied to the highway.

Often, there were noticeable accumulations of chloride at the bottom of the lake during the winter (Fig. 3-20). Apparently the density of either cool tributary water or deep seepage was increased enough

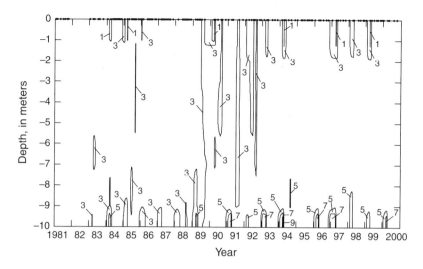

FIGURE 3-20. Isopleths of chloride concentration with depth over time in Mirror Lake.

to flow down-slope and settle at the deepest point of the lake. The turnover of the lake water column in the spring was strong enough to overcome the density differences caused by these introduced ions, and they were mixed thoroughly throughout the lake's water column. The offset seasonal timing and slow increase of fluxes over the first decade in the Northeast Tributary also implied that the salt-affected area was not a simple and fast pass-through system, but it had a capacity to store dissolved road salt, possibly within bedrock fractures or till.

During 1981–2000, the annual volume-weighted concentrations of sodium and chloride rose to a peak around 1996, and then fell off abruptly by about 30 percent the next year, remaining relatively steady or falling slightly from 1997 through 2000. The year 1996 was the wettest of the 20 years (precipitation inputs were greater than 1 standard deviation from the 20-year mean—see chapter 2). The peak in precipitation and runoff during 1996 may have caused a flushing of accumulated road salt from the soil and groundwater reservoirs, resulting in a new equilibrium between salt application and leaching. In 2000–2001, the section of highway adjacent to the lake, and the earthen berm itself, were rebuilt at great effort and expense (post-construction cost estimate was over $500,000 USD) and fitted with an under-liner of impervious plastic sheet designed to prevent salt infiltration (NH-DOT Bureau of Engineering Records: Project No. 13139 Plan for Proposed Safety Improvements and Pavement Rehabilitation I-93, September 2000).

Salt application records for I-93 in the maintenance district servicing the Mirror Lake area were available for 1986–1990 and 1997–2000 (NH-DOT, Bureau of Environment records). There was no significant difference in average road salt application rates between the two periods. Inserting an average value from combining the known rates (19.4 metric tons of salt per lane per km per year), and calculating the mass flux to the original (pre-highway) Northeast Tributary based on the four-lane distance traversed across the watershed (0.6 km), the total estimated highway loading was an impressive 364 metric tons of sodium and 568 metric tons of chloride for 1981–2000. Of this huge potential amount, only 2 percent of the sodium or chloride entered the lake through the truncated Northeast Tributary over 20 years, confirming the results found by Rosenberry et al. (1999). If all the sodium and chloride were to have flowed directly into the lake, and the Northeast Tributary were restored to its previous size of 20 ha (with concomitant increase in discharge), then the concentrations in the NE tributary would have reached about 206 mg L^{-1} sodium and 322 mg L^{-1} Cl^- chloride by 2000, exceeding the U.S. Environmental Protection Agency drinking water standards and chronic aquatic biology toxicity levels for chloride (230 mg L^{-1}; EPA 1988, in Kaushal et al. 2005). If nothing had been done, and considering only inputs from the Northeast Tributary, the lake would have risen to approximately 21 mg L^{-1} sodium and 33 mg L^{-1} chloride in 2000, and would have eventually exceeded the statutory threshold for Cl^- in 2130. Such a deceptively simple back-calculation of environmental risk is only possible because of a full knowledge of each of the mass balance fluxes for the lake.

Since 2000, the concentrations of chloride in the Northeast Tributary have declined gradually by over 30 percent, but average road salt application rates from 2001 to 2007 have actually increased by about 30 percent (NH-DOT, Bureau of Environment records). This pattern suggests the recent reconstruction efforts in the Northeast Tributary were not only successful but absolutely necessary to protect the ecological integrity of the lake.

Surprisingly, the concentrations of sodium and chloride in the lake did not decline in response to the decreases of salt in the Northeast Tributary after 1995, or even after 2000. In fact, from 1995 through 2000, total annual inputs of sodium and chloride to the lake increased by about

34 and 58 percent, respectively (based on annual average values). The primary reason for that increase is because concentrations of sodium-chloride rose in the Northwest and West Tributaries, especially in the period after 1996 (Fig. 3-4a, b, c; Fig. 3-6d and i). During the study, the relatively saline Northeast Tributary carried an annual average of 2 percent of the total water inflow to the lake, verses an average of 35 and 22 percent for the Northwest and West Tributaries, respectively. While chloride concentrations declined in the Northeast Tributary by about 24 mg L^{-1} from a peak of 70 mg L^{-1} in 1995, chloride in the Northwest and West Tributaries increased by 0.8 and 1.3 mg L^{-1}. However, these smaller changes in the Northwest and West Tributary concentrations translated into larger increases in flux due to the greater flow contributions from each tributary. For example, the contribution of the Northeast Tributary to total annual chloride inputs peaked at 50 percent in 1993, when the Northwest and West Tributaries contributed 17 and 20 percent, respectively. By 2000, the annual Northeast Tributary chloride contribution declined to 30 percent, and the Northwest and West Tributaries increased to 24 and 37 percent, respectively.

The most likely cause of these increases in the West and Northwest Tributaries was also deicing salt; in this case, it was applied at much smaller rates to the local town road (Lady Slipper Road) that passes through the Northwest and West Tributaries. Lady Slipper Road was built in 1969 as a concession to the controversy associated with property losses along the I-93 construction corridor. Similar to the Northeast Tributary experience in the 1970s, the Northwest and West Tributary Na:Cl equivalent ratios dropped from >2 in the early 1980s to near the sea salt value of 0.8 by 2000. In this case, while the salt was just a minor (<1% by weight) component of the sand (an "anti-caking" ingredient) used to treat the road and only amounted to an annual average mass input of 440 kg yr^{-1} of chloride to the Northwest and West Watersheds, it had a large effect because of where it was applied. The annual average chloride mass flux to the lake, prior to 1970 for both tributaries combined, is estimated to be about 294 kg yr^{-1} of chloride. The difference (360 kg yr^{-1} Cl) between that pristine condition and the annual average for 1981–2000 (654 kg yr^{-1} Cl) approximates the road salt application rate. This implies that 82 percent of the salt applied to the secondary road is entering the lake through those tributaries.

Plainly, application of the mass balance approach to the existing long-term data has identified and quantified unexpected sources of salt pollution to the lake.

The risk to freshwater aquatic ecosystems by salt pollution is large when the potential mass flux is high (Kaushal et al. 2005). Thus, the greatest concern for Mirror Lake has always been runoff from I-93. However, using knowledge of all the fluxes at the lake, it has been clearly demonstrated that the annual use of about 720 kg yr^{-1} of sodium-chloride on one local road has a greater affect on Mirror Lake currently than the application of an average 19.4 metric tons of NaCl per lane per km per year on nearby I-93. While cause and effect is relatively easy to establish in cases of road salt contamination, the ramifications of changes in mass fluxes are only possible to determine if importance is based on chemical fluxes derived from quantitative measurements of hydrologic and chemical characteristics combined.

CHANGES IN NUTRIENT CONCENTRATIONS AND
BALANCES AND THEIR CAUSES

Are there changes in nutrient concentrations and balances within the lake and what are the causes? Fluxes of the most biologically reactive solutes, ammonium, nitrate, and phosphate, were provided mainly by direct precipitation (82%, 91%, and 55% of total inputs, respectively). Because the 85 ha terrestrial watershed retained about 96 percent of the ammonium, 98 percent of the nitrate, and 83 percent of the phosphate delivered by atmospheric sources, the watershed inputs of these solutes were greatly reduced (Fig. 3-21). Had this strong terrestrial retention not taken place, the annual mass flux to the lake could have been fivefold larger for ammonium, sixfold larger for nitrate, and threefold larger for phosphate. On the other hand, the watershed of Mirror Lake exported large amounts of calcium, magnesium, potassium, sodium, and dissolved silica to the lake through tributary water and ground water (Table 3-8). Except for a portion of the sodium added by road salt (and the effects of ion-exchange; see Discussion topic above), these inputs represented primary or secondary weathering release within the watersheds (S. Bailey et al. 2003).

In sum, the total inputs to the lake over 20 years of the most biologically labile nutrients were 6153 kg of potassium, 1051 kg of ammonium, 6608 kg of nitrate, 135 kg of phosphate, and 84,542 kg of dissolved silica. Of this, 605 kg (10%), 793 kg (75%), 5499 kg (83%), 57 kg (42%), and 56,041 kg (66%), respectively, were retained in the lake. As a means of comparing lake nutrient retention between different solutes over time, we calculated the ratio of the total annual outputs from and total inputs to the lake (1−[O/I], expressed as percent)(Fig. 3-10). This approach also allowed use of the problematic ammonium and phosphate concentration data (see Research Methods, this chapter), because any temporal patterns (e.g., a step-jump) in concentration caused by analytical problems affecting all of the inputs and outputs would be normalized by considering retention ratios rather than trends in actual mass uptake.

The chemical retention trends (Fig. 3-10) indicated that the lake was a strong sink (I > O, or positive retention) for ammonium, nitrate, phosphate, and dissolved silica in each of the 20 years of study. This retention appears consistent with biological utilization of these chemicals within the lake. Except for phosphate, the annual retention was typically above 50 percent for these nutrients. Phosphate retention rose noticeably over the course of the study from near zero in 1981 to above 80 percent in 1995, with an average of 42 percent for the entire period. Ammonium retention climbed steadily from near 50 percent to 85 percent. Nitrate retention increased from 80 to 90 percent overall. Dissolved silica retention ranged from 50 to 80 percent, with no discernable temporal trends.

Our expectation was that essential or limiting nutrients would be strongly retained within the lake. Several mesocosm studies suggested that algal productivity was co-limited by nitrogen and phosphorus in Mirror Lake (Bade et al. 2007; Gerhart and Likens 1975). The lower retention values for phosphate from 1981 to 1991 indicated that nitrogen was more directly limiting. Several plausible situations could cause an increase in retention of a solute. Two reasons are most obvious. First, an increase in the primary limiting nutrient would cause a greater demand for secondary nutrients, increasing retention of the secondary nutrient. Second a decrease in the input of a secondary nutrient would increase its demand and potential for retention. Because the concentrations of nitrogen (as ammonium or nitrate) and phosphorus were often near detection limits, and because of the additional analytical difficulties that were experienced, it was difficult to ascribe an exact mechanism to explain the observed changes. However, it seems evident that there have been significant changes in the retention

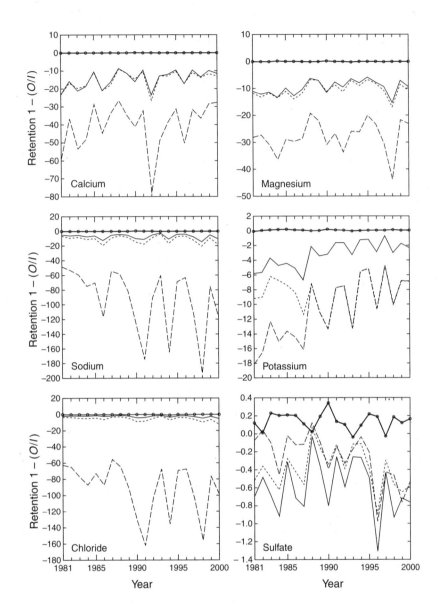

FIGURE 3-21. Annual watershed solute percent retention trends.

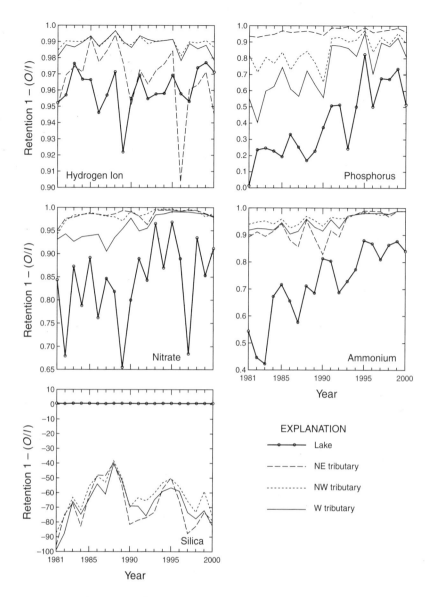

FIGURE 3-21 (CONTINUED).

of nitrogen and phosphorus, and these may indicate a potential shift in the limiting nutrient or trophic status of Mirror Lake.

Non-essential or non-limiting nutrients were not expected to be retained significantly in Mirror Lake. This was the case for the calcium, magnesium, sodium, sulfate, and chloride. Since there was little in-lake processing of these solutes, the changes in concentration observed for some of these solutes must have been driven mainly by external hydrologic forcing (e.g., increases in tributary fluxes of sodium and chloride). However, for some solutes (e.g., calcium, magnesium), there is no observable trend in inputs (Fig. 3-8), but there is a significant decrease in lake concentrations (Fig. 3-6). Apparently dilution is taking place for calcium or magnesium, but the hydrologic variability makes it difficult to discern the ultimate cause. One explanation for the decrease that may apply is that concentrations of the strong acid anion, sulfate, were declining at the same time. Thus, calcium and magnesium were not being leached out of the watershed by acidic deposition as strongly as earlier in the period (Likens et al. 2002). Confounding this situation further, changes in sodium loading to the watershed may be dampening the response of calcium and magnesium to atmospheric inputs.

Dissolved inorganic carbon is a nutrient required for autotrophic productivity; however, it is generally not considered limiting. The strong apparent retention of DIC (from 30% to 50%) represents a combination of biological fixation as well as unmeasured gaseous losses of carbon dioxide (Table 3-7l). In fact, gas exchange was a significant portion of the carbon budget for Mirror Lake during a one-year study by Cole et al. (1994). Diatoms, and to a lesser extent sponges, present in Mirror Lake require silicate. Uptake of silicate therefore depends on habitat and conditions favorable for the growth of diatoms. Diatoms are substantially prone to sedimentation because of the increased density caused by the frustules. A large sedimentary loss of silicate would therefore be expected, and this is evident in 50 to 80 percent of the dissolved silica being retained (Table 3-7n; Fig. 3-10). Although not usually considered an essential nutrient, hydrogen ion is strongly retained (>95%; Fig. 3-10), likely through neutralization by high-ANC ground water (see above).

In summary, Mirror Lake has remained oligotrophic. There did not appear to be an increase in any of the essential nutrients such as nitrate, ammonium, or phosphate, or trends in the stoichiometry of N and P for inputs or the lake pool (Fig. 3-22). Biological productivity is likely

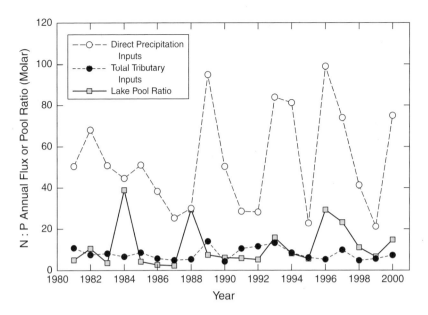

FIGURE 3-22. Nitrogen to phosphorus (molar) ratio trends for all sources of solute input to Mirror Lake.

limited by one or a combination of these solutes. The abundance of sodium and chloride in the lake water was in large excess of what is required for biological functions, and the increased external influx has led to a general increase in concentration. Decreases in the concentration of calcium and magnesium may have been driven by changes in inputs of acidic deposition to the terrestrial watershed, and the subtle influence of that on exports to the lake via tributary water.

GROUNDWATER EFFECTS ON CHEMICAL BUDGETS

Does ground water (in/out) matter to balancing chemical budgets? The importance of ground water to the annual water budget of Mirror Lake can be seen first in the percentage it represents of total annual average water flux. Annual groundwater out (*GWO*) plus groundwater in (*GWI*) composed 34 percent of the total water flux (inputs plus outputs) at Mirror Lake in an average year (chapter 2). Considered separately, *GWI* was 16 percent of total inflow, and *GWO* was 52 percent of outflow, in an average year. For the purposes of estimating mass balances, it was critical to recognize that the largest outflow from Mirror Lake was not, as might be assumed, the surface outlet, but rather seepage to GWO, and that both subsurface fluxes had to be calculated by careful measurement during the study to bring an accurate closure to the budgets. For most solutes, flow of lake water into ground water represented more than 50 percent of the fluvial losses from the lake (see Research Results, this chapter). If groundwater fluxes were ignored in an attempt to calculate mass flux and export of solutes at Mirror Lake, the results would have seriously underestimated the fluxes.

For example, magnesium was considered to be the most likely solute to behave in a conservative manner in Mirror Lake, a fact reflected in the annual input-output balance (I-O). The average magnesium input-output balance was +4 kg yr^{-1}, and this slight retention made sense because magnesium was a component of inorganic sediments in Mirror Lake, and there could have been some small biological uptake (Likens 1985). Magnesium fluxes in GWI and GWO were quite different, averaging +89 and −174 kg yr^{-1}, respectively. However, calculating the magnesium budget based only on precipitation, tributary inflow, and surface outflow mass would have resulted in an imbalance of −85 kg yr^{-1}, or a large release of magnesium from the lake (equal to 42% of the annual streamwater inputs), which is highly unlikely. Sodium had a similar small imbalance (−4 kg yr^{-1} release), but much higher I and O fluxes, and ignorance of the GWI and GWO fluxes could have resulted in highly nonconservative behavior (416 kg yr^{-1} release) which does not make sense for that solute either. Dissolved silica was heavily retained in the lake (2802 kg yr^{-1} on average) by diatom uptake and sedimentation. Overlooking the GWI and GWO average values for dissolved silica would have created an even larger annual imbalance (3469 kg yr^{-1} retention) because in this case the GWO term was much smaller than the GWI term. Estimations of diatom productivity and sequestration founded only on these erroneous values would have been high by at least 20 percent.

Groundwater inflow (GWI) was one of the smallest components of the hydrologic budget, yet GWI was a major contributor to some of the chemical budgets. Average annual GWI flow was calculated to be 112,500 m^3 yr^{-1}, compared to average stream water inputs (417,500 m^3 yr^{-1}) and average GWO seepage (348,896 m^3 yr^{-1}). Thus stream water input and lake seepage to ground water (GWO) were each more than three times the value of flow for groundwater inflow to the lake. Groundwater inflow represented 16 percent of total water input to the lake (chapter 2). Despite this lesser water influx, *GWI* solute fluxes contributed greater than 25 percent of the total annual influx for calcium, magnesium, potassium, sodium, DIC, ANC, and dissolved silica. Groundwater ANC was a major factor in neutralization of direct acidic deposition to the lake, as it provided 63 percent of the annual ANC influx, against 37 percent for all three tributaries. The *GWI* influx of DIC was also the single largest vector (58%) for that solute, and the very large resulting DIC imbalance suggested a high degree of retention in the lake (plant uptake), or a large unmeasured release (possibly as CO_2; Cole et al. 1994). Surprisingly, annual average *GWI* flux of potassium was calculated to be 45 percent of the total inputs to the lake. With the high *GWI* potassium influx, the imbalance in the lake was shifted to retention (35 kg yr^{-1}), which was logical given that potassium is an important plant nutrient.

The primary exception to the seemingly logical imbalance results was chloride, which had a large average annual imbalance of -343 kg yr^{-1} for the 20 years. The sodium flux, which should have roughly matched the chloride flux (4:6 on a weight basis for salt), had an average imbalance near zero (-4.4 kg yr^{-1}). There was a parallel offset in the regression lines describing the sodium and chloride imbalances (as kg yr^{-1}) verses the annual water imbalance, and that offset equaled the chloride imbalance (Fig. 3-23), which certainly implied some systematic problem. This difference between sodium and chloride balances was problematic and suggested four possible explanations: (1) sediment release of stored chloride, (2) incorrect measurement of chloride concentrations, (3) incorrect measurement of chloride flux, and (4) an unmeasured source of chloride. We are aware of no mechanism for sediment release (explanation 1) in a dilute, freshwater lake. Chloride analyses (explanation 2) used highly standardized methods with small error (see Research Methods, this chapter). Arbitrarily doubling the median chloride concentration (1.1 mg L^{-1}) for *GWI* (explanation 2) or doubling the *GWI* water flux (explanation 3) could have balanced the budget. In that case, the *GWI* chloride value would have been far above the 75th percentile (1.3 mg L^{-1}) of all the well data, and the cation-anion arrays would not have balanced. Doubling the *GWI* water flux to compensate (explanation 3) would have created a cascade of large imbalances with all the other solute budgets and would have exceeded the reasonable hydrologic error

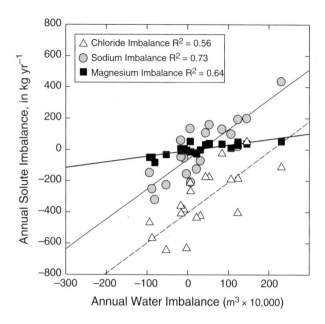

FIGURE 3-23. Annual solute imbalances for magnesium, sodium, and chloride versus annual water imbalances, with regressions (coefficients shown in explanation).

(see chapter 4). Error in the other individual hydrologic or chemical components was not large enough to make up the discrepancy (chapters 4 and 5). When the annual sodium and chloride imbalances were expressed in equivalents versus the annual water imbalances, the slopes of the lines were not parallel, but convergent. This result suggested that chloride was accompanied by a cation other than sodium when the imbalance was positive (retained).

What appeared to be required to solve this imbalance was an unmeasured source of chloride that entered the lake without sodium as the matching cation (explanation 4). The best candidate was perhaps iron-chloride in a reduced form that seeped into the lake via a site-specific groundwater pathway, such as near the Northeast Tributary. Recent ancillary sediment geophysical work supported the existence of a site of high-conductance deep in-seepage offshore of the Northeast Tributary, and apparently nowhere else. The main point of this discussion about a missing chloride source is that by constraining the other hydrologic and chemical fluxes through careful measurements, especially including ground water, the possibilities for alternative, unmeasured pathways can be identified, and the appropriate research conducted to quantify them.

In summary, we have shown several examples where understanding groundwater flow and chemistry is vital to constructing mass budgets of lakes, especially for solutes such as the base cations, DIC, ANC, and dissolved silica, where ground water is a major contributor. Often groundwater flow is calculated by difference in hydrologic budgets, and groundwater chemical flux is considered to be in net balance. A net input or output of ground water has very little utility when constructing a chemical budget. Groundwater "hotspots" can be significant vectors for solutes that otherwise might have poorly constrained net balances. In some instances, the understanding of groundwater contributions is less essential. For example, in Mirror Lake, much of the ammonium, nitrate, and phosphate were supplied from precipitation, and because they were largely retained in the lake, loss via groundwater outflow may have been insignificant. However, without the careful consideration of the potential for groundwater vectors, we could not have made this conclusion.

BIOGEOCHEMICAL INFLUENCE OF WET AND DRY CONDITIONS

What is the biochemical influence of wet and dry conditions? As noted in chapter 2, the study presented in this book was conducted during a relatively wet period in the last century. In this context, what does a wet or dry year mean in terms of the solute budgets for 1981–2000? For solutes supplied mostly by single sources, does it make a difference for the lake?

The three wettest years, on the basis of annual total water input to the lake, were 1981, 1990, and 1996, and for calcium, magnesium, potassium, sodium, sulfate, chloride, and dissolved silica, these were years with the highest influxes, because a high percentage of these solutes are provided via streamflow (Tables 3-6a and 3-11). The three driest years, on the basis of annual total water input to the lake, were 1985, 1988, and 1993, and the same solutes that had high fluxes in the wet years had the smallest fluxes in the dry years. For those solutes, high inputs also meant high outputs, so that the actual input-output balances did not change much. Highs and lows in annual fluxes of ammonium, hydrogen ion, nitrate, and phosphate were not related to wet or dry years, on the basis of annual total water input to the lake (for influence of climate on seasonal patterns see Research Results, this chapter).

As expected, wet and dry periods especially affected the annual streamwater influx data, but not in direct proportion to the change in water flow. Total surface tributary flow in the driest year, 1988, was 64 percent below the total flow of the wettest year, 1996. Solutes derived primarily from the terrestrial watershed, such as the base cations, declined in annual flux by lesser amounts (40 to 50% between the driest and wettest years) than did the annual water flux, because the concentrations of these chemicals had risen abruptly by two- to threefold at the smallest flows, thus partially compensating for the smaller water discharge.

The relative importance of ground water changed dramatically with climate at Mirror Lake (Table 3-12). When the lake pool dropped below the top of the dam, typically in mid-summer, the solute flux for the surface outlet essentially became nil (chapter 2), whereas the GWO flux continued. These dry periods had a significant effect on annual budgets. In the wettest year during 1981 to 2000 (1996), surface outflow (484,000 m^3) accounted for 56 percent of total annual outflow, and GWO (318,000 m^3) made up 37 percent (with 7% evaporation from the lake, 58,800 m^3). In the driest year, based on annual precipitation input (1988), the proportion changed to 17 percent surface outflow (97,600 m^3) and 70 percent GWO (396,000 m^3)(with 13% evaporation, 72,600 m^3). The solute outfluxes followed with similar patterns. In 1996 (the wettest year), the overall ratio of surface outfluxes to GWO fluxes was about 60:40 (evaporation was not a factor in solute flux), and in 1988 (the driest year between 1981 and 2000 based on precipitation input to the lake, and second driest based on the Palmer Hydrologic Drought Index—chapter 2) the same general relationship reversed to 20:80. Failure to account for these differences in surface outflow and GWO together would have seriously compromised all of the annual mass balances in the 20 years of measurement.

The average GWI annual flow was modest and relatively stable (percent coefficient of variation of the mean = 5%), with 12 percent of total water inputs in the wettest year (1996) and 22 percent in the driest year (1988), but there were significant climate-driven changes in solute flux nonetheless. As a general rule, under dry conditions, any potential groundwater storage would have to be replenished by available precipitation inputs before appreciable tributary water flow was generated. So in this case, the reduced tributary flux contributed to a relatively large increase in the importance of *GWI*. For example, in the wettest year of 1996, *GWI* provided 39 percent of the potassium mass flux to the lake, but 52 percent of the mass flux in the driest year of 1988. In 1996, *GWI* added 18 percent of the total sodium inputs, but 34 percent in 1988. The same pattern held for most of the solutes, except for those with very small *GWI* values, such as ammonium, hydrogen ion, nitrate, and phosphate. Thus, in a prolonged dry period, we might expect the lake to reflect greater input of ground water relative to other inputs.

This study included the second wettest year (1996) in the 52-year record at Hubbard Brook Experimental Forest (A. Bailey et al. 2003). In the longer context, including the severe regional droughts in the mid-1960s (up to 31% less annual precipitation than the average; A. Bailey et al. 2003), the *GWI* flux may have been even more important relative to tributary and precipitation inputs. At Mirror Lake, the *GWI* component was not only enriched with base cations and ANC; it was also available as a constant offset to direct acidic precipitation (see first Discussion issue above).

Logically, *GWI* solute fluxes have made the lake less susceptible to the influence of atmospheric acidic deposition by providing increased buffering in the lake. This affect is perhaps especially important in times of drought. Direct annual precipitation to the lake for the driest year between 1981 and 2000 (1988) based on an annual precipitation amount of 136,900 m^3 was 24 percent less than the 20-year mean of 181,000 m^3, whereas annual precipitation in the wettest year (1996), 236,900 m^3, was 31 percent more. Whereas proportion of base cation and dissolved silica flux in direct precipitation had a tendency not to change more than a few percent in these dry or wet years, the proportion ammonium, hydrogen ion, sulfate, and nitrate annual influx increased in the driest year by up to 10 percent. Years with less or more precipitation flux meant much less or much more tributary flux but did not strongly influence *GWI* (Tables 3-11 and 3-12), which varied by only 15 percent between 1988 and 1996.

Lake solute storage was not related significantly to wet or dry years for any solute. The patterns in annual lake solute pools tended to be invariant or dampened in amplitude, compared to the inputs and output fluxes. Exceptions to this uniformity were pools of ammonium, nitrate, and phosphate, which varied two- to threefold between years but were not related to wet or dry conditions. Basically, within the range experienced in this study, and on an annual basis, wet or dry periods changed solute imbalances and lake nutrient storage very little. What were most affected by wet or dry years were the proportions of fluxes from each source, such as GWI and tributary water in or GWO and surface outflow.

CONCLUSIONS

Many characteristics of Mirror Lake have been studied for more than 40 years. As a result, numerous publications have resulted in the lake's becoming highly visible scientifically and becoming a valuable source of information for lake managers. This information, combined with the recent 20 years of detailed monitoring of all components of the lake's water and chemical budgets (for some variables on an hourly basis), provides an unusual opportunity to reach certain conclusions about the effects of climate and human activities on this small New England lake.

The Pervasiveness of Sulfur Sulfate from atmospheric deposition has infiltrated the entire terrestrial and aquatic ecosystem. Groundwater influxes of base cations have made the lake less susceptible to the acidifying influence of atmospheric deposition by providing increased buffering in the lake.

Precipitation Inputs Increasing pH and decreasing sulfate concentration trends in the lake were correlated with similar changes in pH and sulfate fluxes from direct precipitation, and not with changes in any tributary stream.

Neutralization by Groundwater ANC Groundwater ANC was a major factor in neutralization of direct acidic deposition to the lake; it provided 63 percent of the annual ANC influx, against 37 percent for all three tributaries combined.

Increased Acidity Mirror Lake pH would have been between pH 7 and 7.5 prior to input and titration by anthropogenic atmospheric acids. Between 1981 and 2000, the average pH of Mirror Lake was about 6.5, and so it has not returned to what is assumed to be its previous state. The annual volume-weighted average value of pH for the lake between 1967 and 1981 was 6.36 (Likens 1985).

Increasing Sodium and Chloride Over the 20 years of study, the lake solute array shifted subtly from a calcium sulfate–dominated system toward one dominated by sodium and chloride. A return to pristine conditions when the chemistry was a calcium bicarbonate system is unlikely without major reductions in sulfate and salt contamination.

Chloride from Northeast Tributary Although the Northeast Tributary represented just 3 percent of the annual average tributary flow into the lake, by 1994 it contributed about 50 percent of the chloride annual average influx.

Chloride and Sodium from Other Tributaries Concentrations of sodium and chloride in the lake did not decline in response to the decreases of salt in the Northeast Tributary after 1995 because total annual inputs of sodium and chloride to the lake continued to increase. The primary reason for this increase was rising concentrations of sodium chloride in the Northwest and West Tributaries, which contributed 26 percent of total water input to the lake.

Effect of Cation Exchange Equivalent ratios of sodium to chloride in the Northeast Tributary demonstrated the rapid effect of road salt. Ratios dropped quickly from about 3 to 0.8 when the Interstate opened, reflecting road salt sources, and declined further to about 0.4, reflecting sodium ion exchange for calcium and magnesium in the terrestrial watershed. Gradual declines in cation exchange capacity may be responsible for the ratio rising back toward 0.8 since the mid-1980s.

Effect of Nearby Road Salting The use of about 720 kg yr^{-1} of sodium chloride on local roads and driveways within the lake's watershed has had a greater effect on Mirror Lake's current condition than has the application of an average 19.4 metric tons of NaCl per lane per km per year on nearby I-93 since 1970.

Net Retention and Release of Chemicals Data on chemical retention within the lake indicate that the lake retained phosphorus, nitrogen, dissolved silica, and hydrogen ion inputs during 1981–2000.

Retention of ammonium, nitrate, and phosphate in the drainage basin reduced the total influx to the lake to just a fraction of the total atmospheric deposition to the entire watershed.

The pattern in net release or retention for calcium, magnesium, sodium, potassium, and sulfate is primarily related to water movement through the lake. In wet years, when water inputs exceeded water losses, retention values for these constituents in the lake generally approached zero or were negative (released).

The variability in percent annual retention for most solutes was related to changes in streamwater inputs, caused by retention or release in the terrestrial watershed.

Groundwater Inputs and Outputs   Considered separately, GWI was 16 percent of total inflow, and GWO was 52 percent of outflow, for an average year. For the purposes of estimating mass balances, it is critical to recognize that the largest outflow from Mirror Lake was seepage as GWO.

An Unmeasured Cation?   An unmeasured cation (possibly reduced iron) had to have been present to provide the necessary charge to balance the unsupported release of chloride from the lake, and this ion had to have entered the lake through a narrow point-source area undetected by the broader network of sample sites.

Silica   The silica cycle was dominated by large streamwater and groundwater inputs (85 metric tons total) balanced by substantial uptake and/or sedimentation (56 metric tons total), such that about 66 percent of total inputs were sequestered in the sediments during 1981–2000.

Effect of Direct Precipitation on Chemical Budgets   Highs and lows in input-output balances of ammonium, hydrogen ion, nitrate, and phosphate were not related to wet or dry years because these solutes were provided largely (>50%) by direct precipitation.

Groundwater Inputs in Wet and Dry Years   The relative importance of ground water changed with climate at Mirror Lake. In dry years, when direct precipitation, streamwater inputs, and surface water outputs were smallest, the portion of solute mass flux and export by GWI and GWO increased and was maximized.

Water Inputs and Seasonal Chemical Patterns   The modest differences in the distribution of annual water inputs were not enough to create changes in the normal seasonal patterns exhibited by most solutes.

TABLE 3-1. *Methods and precision*

Chemical	Analysis location	Method	POR (%)	MDL
Calcium	IES	FAAS	± 5	0.01 mg L^{-1}
Magnesium	IES	FAAS	± 5	0.01 mg L^{-1}
Sodium	IES	FAAS	± 5	0.01 mg L^{-1}
Potassium	IES	FAAS	± 5	0.01 mg L^{-1}
Ammonium	IES	CFA	± 10	0.01 mg L^{-1}
pH[a]	RSPEL	Potentiometric	± 0.05 units	n/a
Sulfate	IES	IC	± 5	0.02 mg L^{-1}
Nitrate	IES	IC	± 5	0.01 mg L^{-1}
Chloride	IES	IC	± 5	0.02 mg L^{-1}
Phosphate	IES	CFA	± 5	1 µg L^{-1}
Dissolved organic carbon (DOC)	USDA FS (Durham, NH)	AICDD-HTCEC	± 2	0.05 mg L^{-1}
Dissolved inorganic carbon (DIC)[b]	RSPEL	GC	± 5	1 µmol L^{-1}
Acid-neutralizing capacity (ANC)[c]	RSPEL	Gran plot analysis	± 5	1 µeq L^{-1}
Dissolved silica	IES	ICP-AES (reported as SiO$_2$)	± 5	0.1 mg L^{-1}

AICDD-HTCED = analyzed by automated infrared carbon dioxide detection following high-temperature catalyst-enhanced combustion
CFA = continuous flow analysis (automated colorimetry)
FAAS = flame atomic absorption spectrometry
GC = gas chromatography
IC = ion chromatography
ICP-AES = inductively coupled plasma-atomic emission spectroscopy
MDL = method detection limit
POR = precision over range, in %

[a] Measured with combination pH/reference glass electrode, air-equilibrated at lab temperature, and without stirring, within 24 hours of collection.
[b] Analyzed within 24 hours of collection.
[c] Analyzed by auto-titration using 0.1 N hydrochloric acid.

TABLE 3-2. Twenty-year volume-weighted average or median concentrations ($\mu eq\ L^{-1}$) and ion balances for all sources of solute flux and the lake

	Average concentration				Median concentration		Average concentration		
Chemical	Precipitation	Northeast tributary	Northwest tributary	West tributary	Ground water in	Surface water outlet	Ground water out	Lake
Calcium	4.0	411.1	100.2	112.7	196.1	109.4	114.6	120.4
Magnesium	2.0	144.8	34.3	40.4	65.0	38.8	41.0	43.1
Potassium	1.2	32.7	7.4	11.9	31.6	11.3	12.0	12.2
Sodium	4.6	1010.4	57.8	90.4	125.9	87.0	92.5	96.6
Ammonium	13.1	2.2	0.8	1.2	1.1	1.4	1.3	1.3
Hydrogen ion	52.2	4.4	0.9	1.1	0.2	0.8	0.6	0.3
(pH[a])	*(4.28)*	*(5.36)*	*(6.04)*	*(5.96)*	*(6.65)*	*(6.12)*	*(6.23)*	*(6.52)*
Cation sum	77.1	1605.6	201.4	257.6	420.0	248.5	262.0	273.9
Sulfate	42.2	136.0	112.8	107.0	117.1	93.9	94.4	96.3
Nitrate	26.7	1.0	0.7	1.9	0.1	1.8	1.2	0.8
Chloride	6.1	1384.4	31.5	67.4	31.0	76.0	81.2	84.9
Phosphate	0.6	0.4	0.1	0.3	0.1	0.2	0.2	0.2
Bicarbonate (ANC)	0.0	54.2	44.1	59.7	320.0	69.2	73.6	82.9

DOC[b]	5.8	43.6	8.6	17.5	5.3	12.5	11.5	10.5
Anion sum	81.4	1619.7	197.8	253.8	473.6	253.6	262.3	275.6
Total ions	158.5	3225.3	399.2	511.4	893.6	502.1	524.3	549.5
Cation sum − Anion sum[c]	−4.3	−14.1	3.7	3.8	−53.6	−5.1	−0.3	−1.7
Difference between cation sum and anion sum as a percentage of the sum of all ions								
(Cation sum − Anion sum)/Total ions	−2.7%	−0.4%	0.9%	0.7%	−6.9%	−1.0%	−0.05%	−0.3%

[a] pH determined from average or median concentration of the hydrogen ion.
[b] Microequivalents of DOC estimated assuming 1 milligram C per liter equaled 6.0 microequivalents per liter (Buso et al. 2000).
[c] Difference determined from average values in table, not average of difference for individual sample chemical analysis.

TABLE 3-3. *Volume-weighted 20-year average or median concentrations for all sources of solute flux and for Mirror Lake*

Chemical	Units	Precip	NE	NW	West	GWI	Lake	SWO	GWO
		Average	Average	Average	Average	Median[a]	Average	Average	Average
Calcium	Mg L^{-1}	0.08	8.45	2.03	2.29	3.93	2.41	2.20	2.30
Magnesium	Mg L^{-1}	0.02	1.82	0.42	0.50	0.79	0.52	0.47	0.50
Potassium	Mg L^{-1}	0.05	1.30	0.29	0.47	1.24	0.48	0.44	0.47
Sodium	Mg L^{-1}	0.11	23.58	1.34	2.08	2.90	2.22	2.00	2.11
Ammonium	Mg L^{-1}	0.24	0.04	0.01	0.02	0.02	0.02	0.02	0.02
pH[b]		4.28	5.43	6.07	5.99	6.65	6.50	6.17	6.23
Hydrogen ion	μeq L^{-1}	52.8	4.0	0.9	1.1	0.2	0.3	0.7	0.6
Sulfate	Mg L^{-1}	2.04	6.63	5.42	5.18	5.63	4.63	4.49	4.56
Nitrate	Mg L^{-1}	1.67	0.06	0.04	0.13	0.01	0.05	0.12	0.07
Chloride	Mg L^{-1}	0.22	50.03	1.11	2.38	1.10	3.01	2.68	2.86
Phosphate	Mg L^{-1}	0.021	0.012	0.005	0.009	0.003	0.007	0.007	0.007
DOC	Mg L^{-1}	0.95	7.99	1.75	3.10	0.88	1.75	2.02	1.88
DIC	μmol L^{-1}	35	348	93	138	715	136	135	135
Bicarbonate (ANC)[c]	Mg L^{-1}	0	3.4	2.8	3.7	19.5	5.1	4.2	4.2
Dissolved silica	Mg L^{-1}	0.06	11.13	6.51	7.78	11.20	2.21	2.41	2.17

[a] Groundwater values derived from median water table well chemical data.
[b] Average pH determined from average hydrogen ion concentration.
[c] ANC as bicarbonate converted from ANC as μeq L^{-1}.

TABLE 3-4. *Summary of test to determine if annual volume-weighted concentrations in Mirror Lake were representative of a normal distribution*

Chemical	Observations (years)	Shapiro-Wilk statistic (W)	Probability of a smaller value of W	Do data represent a normal distribution? $\alpha = 0.1$
Calcium	20	0.965	0.65	Yes
Magnesium	20	0.958	0.51	Yes
Potassium	20	0.965	0.65	Yes
Sodium	20	0.900	0.04	No
Ammonium	20	0.803	0.001	No
Hydrogen ion	10	0.943	0.59	Yes
Sulfate	20	0.931	0.16	Yes
Nitrate	20	0.888	0.02	No
Chloride	20	0.884	0.02	No
Phosphate	20	0.890	0.02	No
DIC	11	0.929	0.40	Yes
ANC	10	0.908	0.26	Yes
Dissolved silica	20	0.985	0.98	Yes

NOTE: Test statistic is Shapiro-Wilk (W).

TABLE 3-5. STATISTICAL ANALYSES OF THE RELATION BETWEEN ANNUAL VOLUME-WEIGHTED CONCENTRATION AND YEAR FOR MIRROR LAKE

TABLE 3-5A. *Linear regression analysis*

Chemical	Units	Observations (years)	F statistic	P>F	R^2	Is relation significant? $\alpha = 0.01$	Slope
Calcium	Mg L^{-1}	20	47.52	<0.001	0.72	Yes	−0.015
Magnesium	Mg L^{-1}	20	12.74	0.002	0.41	Yes	−0.002
Potassium	Mg L^{-1}	20	11.66	0.003	0.39	Yes	−0.003
Hydrogen ion	µeq L^{-1}	10	5.09	0.054	0.38	No	n/a
Sulfate	Mg L^{-1}	20	100.2	<0.001	0.84	Yes	−0.080
DIC	µmol L^{-1}	11	3.61	0.09	0.26	No	n/a
ANC	µeq L^{-1}	10	0.61	0.45	0.07	No	n/a
Dissolved silica	Mg L^{-1}	20	0.04	0.83	0.002	No	n/a

TABLE 3-5B. *Kendall tau correlation coefficient analysis*

Chemical	Units	Observations (years)	Kendall tau r statistic	P>!r!	Is relation significant? $\alpha = 0.01$	Trend direction
Sodium	Mg L^{-1}	20	0.89	<0.001	Yes	Increasing
Ammonium	Mg L^{-1}	20	−0.51	0.0016	Yes	Decreasing
Nitrate	Mg L^{-1}	20	−0.008	0.62	No	None
Chloride	Mg L^{-1}	20	0.86	<0.001	Yes	Increasing
Phosphate	Mg L^{-1}	20	−0.57	0.004	Yes	Decreasing

TABLE 3-6. *Summary of selected statistical parameters examining the relation between annual volume-weighted concentration and year for atmospheric precipitation, the tributaries, and the surface outlet of the lake*

Chemical	Precipitation		Northeast tributary		Northwest tributary		West tributary		SWO		GWO	
	R^2 or (r)	Trend	R^2	Trend	R^2 or (r)	Trend	R^2 or (r)	Trend	R^2 or (r)	Trend	R^2 or (r)	Trend
Calcium	0.09	None	0.03	None	0.66	Decreasing	0.23	Decreasing	0.27	Decreasing	0.42	Decreasing
Magnesium	0.03	None	0.01	None	0.46	Decreasing	0.16	None	0.22	Decreasing	0.30	Decreasing
Potassium	0.49	Increasing	0.00	None	(0.02)	None	0.00	None	0.05	None	0.14	None
Sodium	0.03	None	0.61	Increasing	0.76	Increasing	0.80	Increasing	0.71	Increasing	(0.72)	Increasing
Ammonium	0.38	Increasing	0.39	Decreasing	(−0.52)	Decreasing	(−0.50)	Decreasing	0.70	Decreasing	0.81	Decreasing
Hydrogen ion	0.33	Decreasing	0.03	None	0.07	None	0.00	None	(−0.33)	Decreasing	0.26	Decreasing
Sulfate	(−0.36)	Decreasing	0.03	None	0.91	Decreasing	0.61	Decreasing	0.77	Decreasing	0.88	Decreasing
Nitrate	(0.01)	None	0.30	Decreasing	(−0.35)	Decreasing	(−0.54)	Decreasing	(−0.37)	Decreasing	0.13	None
Chloride	(0.11)	None	0.36	Increasing	0.79	Increasing	(0.74)	Increasing	0.72	Increasing	0.83	Decreasing
Phosphate	(−0.07)	None	0.37	Decreasing	(−0.47)	Decreasing	(−0.58)	Decreasing	(−0.55)	Decreasing	(−0.55)	Decreasing
DOC	0.01	None	0.91	Increasing	0.10	None	0.21	None	(−0.04)	None	0.00	None
DIC	0.02	None	0.05	None	0.12	None	0.19	None	0.06	None	0.17	None
ANC	0	None	0.08	None	0.06	None	0.04	None	0.00	None	(−0.33)	None
Dissolved silica	(−0.04)	None	0.01	None	0.17	None	0.09	None	0.03	None	0.00	None

NOTE: Statistical summary includes simple linear regression coefficient (R^2) for solutes that did represent a normal distribution based on the Shapiro-Wilk test at $\alpha = 0.1$, as well as Kendall tau correlation coefficient (r) for solutes that did not represent a normal distribution based on the Shapiro-Wilk test at $\alpha = 0.1$. The relation between annual volume-weighted average concentration and year was considered significant if the test statistic (probability of a greater F or absolute value of r) was less than 0.05—the 95% confidence level.

TABLE 3-7. ANNUAL INFLUX TO AND OUTFLUX FROM MIRROR LAKE

TABLE 3-7A. *Twenty-year average annual inputs to Mirror Lake: Mass, volume, and percent (%) of total inputs*

Chemical	Units yr^{-1}	Inputs total	Precip in	Precip in %total	NE in	NE in %total	NW in
Calcium	kg	1410.2	14.4	1.0%	100.0	7.1%	498.2
Magnesium	kg	295.3	4.4	1.5%	21.4	7.2%	103.4
Potassium	kg	307.6	8.6	2.8%	15.5	5.0%	71.5
Sodium	kg	1283.6	19.3	1.5%	282.1	22.0%	329.6
Ammonium	kg	52.5	42.85	81.6%	0.49	0.9%	3.65
Hydrogen ion	eq	9920.3	9444.3	95.2%	52.9	0.5%	223.9
Sulfate	kg	3234.3	367.2	11.4%	79.4	2.5%	1344.9
Nitrate	kg	330.4	299.51	90.7%	0.76	0.2%	10.55
Chloride	kg	1411.9	39.5	2.8%	596.2	42.2%	276.7
Phosphate	kg	6.7	3.69	54.8%	0.14	2.1%	1.17
DOC	kg	1210.8	172.0	14.2%	88.2	7.3%	393.6
DIC	mol	135,393.3	6400	4.7%	4329	3.2%	23,237
ANC	eq	56,992.1	0	0.0%	658	1.2%	10,931
Dissolved silica	kg	4227.1	11.2	0.3%	134.2	3.2%	1605.6
Water	m^3	711,080	181,098	25.5%	12,140	1.7%	248,049

NW in % total	West in	West in % total	GWI	GWI % total	Qc/Qa (% check) sum	SWI total	SWI % total
35.3%	355.2	25.2%	442.3	31.4%	100.0	953.4	67.6%
35.0%	77.2	26.1%	88.9	30.1%	100.0	202.0	68.4%
23.2%	73.1	23.8%	139.0	45.2%	100.0	160.1	52.0%
25.7%	326.8	25.5%	325.8	25.4%	100.0	938.5	73.1%
7.0%	3.28	6.3%	2.25	4.3%	100.0	7.4	14.1%
2.3%	173.9	1.8%	25.2	0.3%	100.0	450.8	4.5%
41.6%	809.1	25.0%	633.6	19.6%	100.0	2233.5	69.1%
3.2%	18.66	5.6%	0.90	0.3%	100.0	30.0	9.1%
19.6%	375.8	26.6%	123.8	8.8%	100.0	1248.6	88.4%
17.3%	1.40	20.8%	0.34	5.0%	100.0	2.7	40.2%
32.5%	457.8	37.8%	99.1	8.2%	100.0	939.6	77.6%
17.2%	22,206	16.4%	79,221	58.5%	100.0	49,772.3	36.8%
19.2%	9390	16.5%	36,013	63.2%	100.0	20,979.1	36.8%
38.0%	1215.7	28.8%	1260.5	29.8%	100.0	2955.4	69.9%
34.9%	157,271	22.1%	112,522	15.8%	100.0	417,459.6	58.7%

TABLE 3-7B. *Twenty-year average annual solute outputs from Mirror Lake: Mass export and percent (%) of total outputs*

Chemical	Units	Outputs	GWO	GWO
	yr^{-1}	total		%total
Calcium	kg	1365.9	801.1	58.6%
Magnesium	kg	295.2	173.9	58.9%
Potassium	kg	277.4	164.2	59.2%
Sodium	kg	1256.5	741.7	59.0%
Ammonium	kg	14.2	7.9	55.6%
Hydrogen ion	eq	398.5	204.8	51.4%
Sulfate	kg	2746.6	1583.9	57.7%
Nitrate	kg	54.8	26.0	47.5%
Chloride	kg	1698.6	1004.5	59.1%
Phosphate	kg	4.1	2.4	59.0%
DOC	kg	1207.1	670.8	55.6%
DIC	mol	85,660.1	48,721.6	56.9%
ANC	eq	43,565.0	25,732.3	59.1%
Dissolved silica	kg	1379.7	758.5	55.0%
Water	m^3	680,264.5	348,896.5	51.3%
Evaporation	m^3	73,856.0		

[a] Evaporation = 10.8%.

SWO	SWO %total	I-O mass	I-O %total	Qc/Qa (%check) sum
564.9	41.4%	236.2	17.3	100.0%
121.3	41.1%	52.7	17.8	100.0%
113.2	40.8%	50.9	18.4	100.0%
514.8	41.0%	226.9	18.1	100.0%
6.3	44.4%	1.6	11.3	100.0%
193.7	48.6%	11.2	2.8	100.0%
1162.7	42.3%	421.2	15.3	100.0%
28.8	52.5%	−2.8	−5.1	100.0%
694.1	40.9%	310.5	18.3	100.0%
1.7	41.0%	0.7	18.0	100.0%
536.3	44.4%	134.5	11.1	100.0%
36,938.4	43.1%	11,783.2	13.8	100.0%
17,832.7	40.9%	7899.7	18.1	100.0%
621.2	45.0%	137.3	9.9	100.0%
257,512.0	37.9%	91,384.5	13.4	89.1%[a]

TABLE 3-8. ANNUAL SOLUTE FLUX, SOLUTE BUDGETS, RETENTION, AND BUDGET RESIDUALS FOR ALL 20 YEARS

TABLE 3-8A. *Annual calcium budget, in kilograms/year*

Calendar year	P	NE in	NW in	West in	SWI	GWI
1981	13.4	139.9	752.4	520.9	1413.2	454.1
1982	11.7	76.1	453.4	326.9	856.4	453.4
1983	10.8	102.1	550.8	361.6	1014.5	454.3
1984	13.6	116.0	618.9	418.0	1152.9	475.7
1985	15.4	79.4	401.8	289.0	770.2	445.5
1986	11.2	88.8	570.0	394.4	1053.2	454.9
1987	10.6	64.0	424.7	322.5	811.2	455.9
1988	15.5	74.3	345.6	244.2	664.1	409.0
1989	18.2	112.9	524.1	355.4	992.4	412.5
1990	18.9	138.9	750.5	481.0	1370.4	469.9
1991	19.8	114.5	487.4	357.9	959.8	456.5
1992	8.1	111.4	458.2	352.6	922.2	432.8
1993	12.2	105.3	386.4	272.4	764.1	423.2
1994	14.0	96.2	426.6	316.5	839.3	432.1
1995	17.4	97.0	424.1	312.0	833.1	392.6
1996	15.9	141.3	661.0	466.5	1268.8	475.1
1997	16.0	90.9	393.8	302.8	787.5	465.3
1998	14.4	93.5	485.1	341.5	920.1	428.7
1999	16.5	84.1	414.7	334.5	833.3	416.4
2000	15.0	74.2	434.9	333.1	842.2	437.8
Calcium budget statistical summary						
Average	14.4	100.0	498.2	355.2	953.4	442.3
±1 SD	3.0	22.7	118.1	70.3	206.0	23.3
%CV	21.1	22.6	23.7	19.8	21.6	5.3
Sums	288.8	2000.8	9964.4	7103.7	19,068.9	8845.7
Max	19.8	141.3	752.4	520.9	1413.2	475.7
Min	8.1	64.0	345.6	244.2	664.1	392.6
Median	14.7	96.6	455.8	338.0	888.2	449.4

Total in	SWO	GWO	Total out	Net balance I-O	Retention =1−(O/I) %	Pool change ΔS	Residual I-O-ΔS
1880.7	1078.3	763.2	1841.5	39.2	2.1%	−287.7	326.9
1321.5	484.3	841.9	1326.2	−4.6	−0.3%	31.6	−36.2
1479.6	565.2	717.3	1282.5	197.1	13.3%	−65.9	263.0
1642.3	659.4	691.2	1350.6	291.6	17.8%	91.3	200.4
1231.0	348.2	702.6	1050.8	180.2	14.6%	−108.8	289.0
1519.3	564.9	732.9	1297.8	221.4	14.6%	−30.5	252.0
1277.7	358.1	865.9	1224.0	53.6	4.2%	−2.2	55.8
1088.6	219.3	952.2	1171.5	−82.9	−7.6%	48.1	−131.0
1423.1	536.4	832.6	1369.0	54.1	3.8%	−228.8	282.9
1859.3	768.7	681.0	1449.7	409.5	22.0%	210.4	199.1
1436.1	564.1	901.4	1465.5	−29.4	−2.1%	−13.1	−16.4
1363.1	573.3	842.8	1416.1	−53.0	−3.9%	−59.3	6.2
1199.5	499.6	909.4	1409.0	−209.5	−17.5%	53.1	−262.6
1285.4	492.6	870.0	1362.6	−77.2	−6.0%	33.5	−110.6
1243.1	467.3	772.4	1239.7	3.5	0.3%	−71.3	74.7
1759.7	1017.7	667.8	1685.5	74.2	4.2%	−135.7	209.9
1268.9	539.0	867.1	1406.1	−137.3	−10.8%	−3.4	−133.9
1363.2	532.8	761.9	1294.7	68.5	5.0%	159.6	−91.1
1266.2	460.5	880.8	1341.3	−75.0	−5.9%	−190.8	115.8
1295.0	567.5	767.1	1334.6	−39.6	−3.1%	112.7	−152.3
Calcium budget statistical summary							
1410.2	564.9	801.1	1365.9	44.2	2.2%	−22.9	67.1
220.5	200.9	85.6	169.3	151.8	10.2%	126.4	178.9
15.6	35.6	10.7	12.4	n/a	n/a	n/a	n/a
28,203.4	11,297.2	16,021.5	27,318.7	n/a	n/a	n/a	n/a
1880.7	1078.3	952.2	1841.5	409.5	22.0%	210.4	326.9
1088.6	219.3	667.8	1050.8	−209.5	−17.5%	−287.7	−262.6
1342.3	537.7	802.5	1345.9	21.3	1.2%	−8.2	65.3

TABLE 3-8B. *Annual magnesium budget, in kilograms/year*

Calendar year	P	NE in	NW in	West in	SWI	GWI
1981	5.5	27.8	152.1	111.8	291.7	91.3
1982	3.2	15.7	97.0	72.5	185.2	91.1
1983	3.9	21.5	110.8	78.7	211.0	91.3
1984	3.7	23.9	121.9	85.4	231.2	95.6
1985	3.2	16.6	80.0	61.6	158.2	89.5
1986	3.6	19.1	114.2	85.8	219.1	91.4
1987	3.4	17.5	89.0	70.0	176.5	91.6
1988	4.5	15.7	75.1	54.1	144.9	82.2
1989	5.9	23.4	107.9	76.9	208.2	82.9
1990	5.3	29.2	153.1	102.4	284.7	94.5
1991	5.2	24.9	103.9	78.5	207.3	91.8
1992	4.1	24.7	99.9	79.9	204.5	87.0
1993	4.9	22.9	82.6	61.2	166.7	85.1
1994	4.3	20.2	88.5	68.1	176.8	86.9
1995	5.6	20.3	88.0	68.3	176.6	78.9
1996	7.0	29.5	140.8	101.6	271.9	95.5
1997	3.6	19.7	86.2	67.6	173.5	93.5
1998	2.6	20.2	98.1	73.6	191.9	86.2
1999	4.7	18.5	86.5	72.0	177.0	83.7
2000	3.9	16.2	92.9	73.1	182.2	88.0
Magnesium budget statistical summary						
Average	4.4	21.4	103.4	77.2	202.0	88.9
±1 SD	1.1	4.3	22.9	14.5	40.8	4.7
%CV	24.8	20.2	22.2	18.8	20.2	5.3
Sums	88.2	427.5	2068.5	1543.1	4039.1	1778.1
Max	7.0	29.5	153.1	111.8	291.7	95.6
Min	2.6	15.7	75.1	54.1	144.9	78.9
Median	4.2	20.2	97.5	73.3	188.5	90.3

Total in	SWO	GWO	Total out	Net balance I-O	Retention =1−(O/I) %	Pool change ΔS	Residual I-O-ΔS
388.4	231.7	162.4	394.1	−5.6	−1.4%	−41.9	36.3
279.5	102.8	179.5	282.3	−2.8	−1.0%	−9.3	6.5
306.2	119.2	154.0	273.2	33.0	10.8%	−16.6	49.6
330.6	135.8	145.9	281.7	48.9	14.8%	18.3	30.6
250.9	73.7	149.0	222.7	28.3	11.3%	−11.5	39.7
314.2	123.7	162.4	286.1	28.1	8.9%	18.1	9.9
271.6	78.4	192.2	270.6	1.0	0.4%	8.2	−7.1
231.6	48.6	216.6	265.2	−33.6	−14.5%	16.4	−50.0
297.0	118.1	185.1	303.2	−6.2	−2.1%	−59.5	53.2
384.5	164.8	147.0	311.8	72.6	18.9%	19.8	52.9
304.3	119.3	191.0	310.3	−6.0	−2.0%	7.7	−13.7
295.6	124.4	183.2	307.6	−12.0	−4.1%	−8.4	−3.7
256.7	108.8	199.8	308.6	−52.0	−20.2%	−1.6	−50.3
267.9	103.5	187.7	291.2	−23.2	−8.7%	8.4	−31.6
261.1	102.5	169.0	271.5	−10.4	−4.0%	−9.1	−1.3
374.4	222.0	146.2	368.2	6.2	1.7%	−32.1	38.4
270.7	116.0	188.7	304.7	−34.0	−12.6%	49.2	−83.2
280.7	112.1	161.4	273.5	7.2	2.6%	−25.0	32.1
265.4	97.4	188.1	285.5	−20.0	−7.5%	−26.1	6.1
274.1	122.6	169.4	292.0	−17.9	−6.5%	8.7	−26.6
Magnesium budget statistical summary							
295.3	121.3	173.9	295.2	0.1	−0.8%	−4.3	4.4
44.2	43.4	20.2	36.2	30.0	10.0%	25.1	38.5
15.0	35.7	11.6	12.3	n/a	n/a	n/a	n/a
5905.4	2425.4	3478.5	5903.9	n/a	n/a	n/a	n/a
388.4	231.7	216.6	394.1	72.6	18.9%	49.2	53.2
231.6	48.6	145.9	222.7	−52.0	−20.2%	−59.5	−83.2
280.1	117.1	174.5	288.6	−5.8	−1.7%	−5.0	6.3

TABLE 3-8C. *Annual potassium budget, in kilograms/year*

Calendar year	P	NE in	NW in	West in	SWI	GWI
1981	6.3	20.9	98.3	102.6	221.8	142.7
1982	4.3	13.1	65.9	68.9	147.9	142.5
1983	6.5	15.0	70.0	74.4	159.4	142.8
1984	5.9	16.4	77.2	73.1	166.7	149.5
1985	4.3	10.9	53.6	57.8	122.3	140.0
1986	5.0	13.3	69.9	74.4	157.6	142.9
1987	3.4	10.1	60.3	67.1	137.5	143.3
1988	7.4	10.5	52.8	50.0	113.3	128.5
1989	7.6	15.9	77.8	77.1	170.8	129.6
1990	10.0	24.8	96.6	96.2	217.6	147.7
1991	11.2	17.0	67.5	71.5	156.0	143.5
1992	11.0	16.2	67.0	76.1	159.3	136.0
1993	6.1	15.1	60.2	59.1	134.4	133.0
1994	12.4	14.0	63.6	63.5	141.1	135.8
1995	12.8	13.6	63.8	64.6	142.0	123.4
1996	11.7	23.6	104.1	99.4	227.1	149.3
1997	16.4	16.2	65.0	66.0	147.2	146.2
1998	8.1	15.5	74.3	70.4	160.2	134.7
1999	10.4	14.1	65.0	70.8	149.9	130.8
2000	10.2	13.9	77.6	78.5	170.0	137.6
	Potassium budget statistical summary					
Average	8.6	15.5	71.5	73.1	160.1	139.0
±1 SD	3.4	3.9	14.0	13.4	30.6	7.3
%CV	40.2	24.9	19.6	18.3	19.1	5.3
Sums	171.0	310.1	1430.5	1461.5	3202.1	2779.8
Max	16.4	24.8	104.1	102.6	227.1	149.5
Min	3.4	10.1	52.8	50.0	113.3	123.4
Median	7.8	15.1	67.2	71.2	156.8	141.2

Total in	SWO	GWO	Total out	Net balance I-O	Retention =1−(O/I) %	Pool change ΔS	Residual I-O-ΔS
370.8	231.4	161.8	393.2	−22.4	−6.1%	−67.9	45.5
294.7	91.4	175.1	266.5	28.2	9.6%	−9.3	37.5
308.7	111.3	146.8	258.1	50.6	16.4%	−8.0	58.6
322.1	118.4	133.3	251.7	70.4	21.9%	9.6	60.9
266.6	67.7	139.1	206.8	59.8	22.4%	−20.0	79.8
305.5	108.9	145.1	254.1	51.5	16.8%	−7.9	59.3
284.2	70.8	175.3	246.1	38.1	13.4%	34.1	4.0
249.2	45.0	206.0	251.0	−1.8	−0.7%	33.7	−35.5
308.0	117.8	181.9	299.7	8.3	2.7%	−76.8	85.1
375.2	152.5	135.7	288.3	87.0	23.2%	−23.9	110.8
310.6	103.4	166.8	270.2	40.5	13.0%	−9.4	49.9
306.3	112.5	166.2	278.7	27.7	9.0%	8.9	18.7
273.5	99.7	186.1	285.8	−12.3	−4.5%	7.3	−19.6
289.3	94.2	179.4	273.6	15.8	5.4%	17.1	−1.3
278.2	98.5	161.2	259.7	18.5	6.7%	16.8	1.7
388.1	217.3	143.3	360.6	27.5	7.1%	−23.7	51.2
309.8	105.7	178.6	284.3	25.5	8.2%	23.4	2.0
303.1	104.5	153.7	258.2	44.9	14.8%	−16.4	61.3
291.1	92.5	178.7	271.2	19.9	6.8%	−8.8	28.7
317.8	121.4	169.1	290.4	27.3	8.6%	26.0	1.4
Potassium budget statistical summary							
307.6	113.2	164.2	277.4	30.2	9.7%	−4.7	35.0
35.4	44.1	19.3	40.0	26.7	8.2%	29.5	37.9
11.5	38.9	11.8	14.4	n/a	n/a	n/a	n/a
6152.9	2264.9	3283.1	5548.0	n/a	n/a	n/a	n/a
388.1	231.4	206.0	393.2	87.0	23.2%	34.1	110.8
249.2	45.0	133.3	206.8	−22.4	−6.1%	−76.8	−35.5
305.9	105.1	166.5	270.7	27.6	8.8%	−7.9	41.5

TABLE 3-8D. *Annual sodium budget, in kilograms/year*

Calendar year	P	NE in	NW in	West in	SWI	GWI
1981	29.6	253.6	397.4	388.7	1039.7	334.5
1982	14.2	132.3	244.9	232.8	610.0	334.0
1983	18.6	193.6	286.5	272.3	752.4	334.7
1984	16.4	215.3	331.1	315.5	861.9	350.4
1985	12.8	157.8	219.0	216.0	592.8	328.1
1986	9.9	201.3	322.2	314.9	838.4	335.1
1987	16.4	157.3	254.1	250.5	661.9	335.8
1988	17.2	175.5	209.0	182.8	567.3	301.3
1989	22.1	314.1	339.0	305.9	959.0	303.8
1990	19.7	436.9	510.6	492.7	1440.2	346.2
1991	11.5	349.7	332.4	335.1	1017.2	336.3
1992	21.7	345.3	316.0	334.9	996.2	318.8
1993	30.7	324.9	258.7	248.1	831.7	311.8
1994	10.7	306.3	295.0	295.0	896.3	318.3
1995	24.5	295.3	312.9	314.4	922.6	289.2
1996	44.7	495.1	528.6	519.3	1543.0	349.9
1997	15.6	311.3	320.9	323.3	955.5	342.8
1998	10.1	340.1	368.8	346.4	1055.3	315.8
1999	25.2	330.8	346.8	396.9	1074.5	306.7
2000	15.0	305.2	397.7	450.3	1153.2	322.5
Sodium budget statistical summary						
Average	19.3	282.1	329.6	326.8	938.5	325.8
±1 SD	8.5	95.4	83.3	88.2	253.6	17.2
%CV	44.2	33.8	25.3	27.0	27.0	5.3
Sums	386.7	5641.7	6591.6	6535.8	18,769.1	6516.1
Max	44.7	495.1	528.6	519.3	1543.0	350.4
Min	9.9	132.3	209.0	182.8	567.3	289.2
Median	16.8	305.8	321.5	315.2	939.1	331.1

Total in	SWO	GWO	Total out	Net balance I-O	Retention −1 (O/I) %	Pool change ΔS	Residual I-O-ΔS
1403.8	789.4	548.9	1338.3	65.5	4.7%	−64.7	130.2
958.2	343.5	619.4	962.9	−4.7	−0.5%	49.7	−54.4
1105.6	420.4	541.5	961.9	143.8	13.0%	−49.4	193.2
1228.8	498.7	534.3	1033.1	195.7	15.9%	177.0	18.8
933.8	276.0	562.2	838.2	95.6	10.2%	−62.7	158.3
1183.4	448.3	598.4	1046.7	136.7	11.5%	37.8	98.8
1014.1	297.7	730.7	1028.4	−14.3	−1.4%	32.7	−47.0
885.8	185.4	817.3	1002.7	−116.9	−13.2%	134.9	−251.8
1285.0	484.2	748.9	1233.1	51.9	4.0%	−82.4	134.3
1806.1	733.6	644.5	1378.1	427.9	23.7%	−7.2	435.2
1365.0	521.4	840.7	1362.1	2.9	0.2%	213.0	−210.1
1336.7	561.8	826.0	1387.8	−51.1	−3.8%	105.5	−156.6
1174.2	497.5	928.4	1425.9	−251.7	−21.4%	−102.8	−148.9
1225.4	476.6	890.4	1367.0	−141.6	−11.6%	85.3	−227.0
1236.3	480.7	795.9	1276.6	−40.3	−3.3%	32.3	−72.6
1937.6	1094.4	718.9	1813.3	124.4	6.4%	−74.5	198.8
1313.9	550.5	916.5	1467.0	−153.2	−11.7%	168.2	−321.4
1381.3	532.9	779.9	1312.8	68.4	5.0%	−38.5	107.0
1406.4	470.7	911.7	1382.4	23.9	1.7%	−35.3	59.3
1490.7	632.5	879.7	1512.2	−21.5	−1.4%	104.1	−125.6
		Sodium budget statistical summary					
1283.6	514.8	741.7	1256.5	27.1	1.4%	31.1	−4.1
263.0	197.1	137.8	239.4	146.6	10.8%	96.1	190.1
20.5	38.3	18.6	19.1	n/a	n/a	n/a	n/a
25,671.9	10,296.2	14,834.3	25,130.6	n/a	n/a	n/a	n/a
1937.6	1094.4	928.4	1813.3	427.9	23.7%	213.0	435.2
885.8	185.4	534.3	838.2	−251.7	−21.4%	−102.8	−321.4
1260.7	490.9	764.4	1325.6	13.4	1.0%	32.5	−14.1

TABLE 3-8E. *Annual ammonium budget, in kilograms/year*

Calendar year	P	NE in	NW in	West in	SWI	GWI
1981	48.3	0.9	6.7	6.3	13.9	2.3
1982	32.3	0.5	3.9	3.9	8.3	2.3
1983	32.9	0.6	3.6	4.1	8.3	2.3
1984	33.0	0.5	4.5	4.3	9.3	2.4
1985	29.8	0.3	2.8	2.6	5.7	2.3
1986	34.1	0.7	5.8	5.2	11.7	2.3
1987	28.0	0.7	4.0	3.9	8.6	2.3
1988	38.3	0.3	2.7	2.5	5.5	2.1
1989	48.0	0.9	5.1	5.3	11.3	2.1
1990	53.3	1.6	8.0	7.2	16.9	2.4
1991	56.0	0.8	4.3	3.8	9.0	2.3
1992	43.3	0.8	3.8	5.1	9.8	2.2
1993	38.7	0.2	3.1	2.0	5.3	2.2
1994	44.9	0.2	2.4	1.6	4.2	2.2
1995	57.7	0.1	2.0	1.7	3.8	2.0
1996	56.4	0.1	2.5	1.7	4.3	2.4
1997	49.7	0.1	3.0	1.4	4.6	2.4
1998	37.0	0.2	2.0	1.4	3.6	2.2
1999	48.0	0.1	1.3	0.8	2.2	2.1
2000	47.4	0.1	1.3	0.8	2.2	2.2
Ammonium budget statistical summary						
Average	42.8	0.5	3.7	3.3	7.4	2.3
±1 SD	9.4	0.4	1.8	1.9	4.0	0.1
%CV	22.0	81.0	48.0	57.7	53.5	5.3
Sums	857.0	9.8	73.1	65.7	148.5	45.0
Max	57.7	1.6	8.0	7.2	16.9	2.4
Min	28.0	0.1	1.3	0.8	2.2	2.0
Median	44.1	0.4	3.4	3.2	7.0	2.3

Total in	SWO	GWO	Total out	Net balance I-O	Retention =1−(O/I) %	Pool change ΔS	Residual I-O-ΔS
64.5	17.7	11.7	29.4	35.1	54.4%	−27.6	62.6
42.9	10.1	13.7	23.8	19.1	44.5%	30.2	−11.1
43.5	13.5	11.6	25.1	18.4	42.3%	−51.9	70.3
44.7	7.7	7.1	14.8	29.9	67.0%	−0.8	30.8
37.8	3.4	7.3	10.6	27.1	71.8%	8.6	18.6
48.1	7.2	9.4	16.6	31.5	65.4%	−6.0	37.5
38.9	5.9	10.6	16.5	22.4	57.6%	12.9	9.5
45.9	3.1	10.2	13.2	32.7	71.2%	6.0	26.7
61.4	9.2	10.1	19.3	42.1	68.6%	−12.0	54.1
72.6	7.2	6.3	13.5	59.1	81.4%	−6.0	65.1
67.3	5.1	8.1	13.2	54.1	80.4%	−6.1	60.2
55.3	7.2	10.1	17.2	38.1	68.8%	0.9	37.2
46.1	4.3	8.3	12.6	33.5	72.7%	6.9	26.6
51.3	4.9	6.8	11.6	39.7	77.4%	−7.8	47.5
63.5	2.9	4.7	7.6	55.9	88.0%	−1.7	57.7
63.1	4.9	3.4	8.3	54.9	86.9%	4.4	50.5
56.7	3.5	7.3	10.8	45.9	80.9%	−4.4	50.3
42.8	2.7	3.2	6.0	36.8	86.1%	6.1	30.8
52.3	2.2	4.2	6.4	46.0	87.8%	−9.5	55.5
51.8	3.9	4.4	8.3	43.5	84.0%	30.2	13.3
		Ammonium budget statistical summary					
52.5	6.3	7.9	14.2	38.3	71.9%	−1.4	39.7
10.1	3.9	3.0	6.4	12.0	13.7%	17.8	21.5
19.2	62.1	37.8	44.6	n/a	n/a	n/a	n/a
1050.5	126.3	158.5	284.8	n/a	n/a	n/a	n/a
72.6	17.7	13.7	29.4	59.1	88.0%	30.2	70.3
37.8	2.2	3.2	6.0	18.4	42.3%	−51.9	−11.1
51.6	5.0	7.7	13.2	37.4	72.2%	−1.3	42.5

TABLE 3-8F. *Annual hydrogen ion budget, in eq/year*

Calendar year	P	NE in	NW in	West in	SWI	GWI
1981	12,536.4	107.3	435.0	380.4	922.7	25.9
1982	9747.3	51.5	214.3	180.8	446.6	25.8
1983	9631.9	41.9	219.2	204.3	465.4	25.9
1984	9719.6	48.2	219.0	163.0	430.2	27.1
1985	9488.6	12.0	142.8	95.6	250.4	25.4
1986	10,280.4	39.8	310.8	204.8	555.4	25.9
1987	8219.2	22.3	161.8	107.4	291.5	26.0
1988	9820.7	9.6	67.4	45.5	122.5	23.3
1989	10,859.0	43.5	276.5	168.2	488.2	23.5
1990	12,747.5	107.2	345.4	276.9	729.5	26.8
1991	10,376.8	49.3	162.5	104.1	315.9	26.0
1992	8142.6	50.8	207.6	109.7	368.1	24.7
1993	8787.3	42.5	194.8	139.1	376.4	24.1
1994	9395.9	38.5	198.9	136.5	373.9	24.6
1995	9134.9	25.2	175.0	124.6	324.8	22.4
1996	9173.2	153.4	383.8	313.2	850.4	27.1
1997	8511.7	58.4	195.1	150.9	404.4	26.5
1998	7669.2	48.8	185.9	172.3	407.0	24.4
1999	6599.8	32.6	136.6	127.0	296.2	23.7
2000	8043.9	76.0	245.1	274.6	595.7	25.0
Hydrogen ion budget statistics for 1991 to 2000						
Average	8583.5	57.5	208.5	165.2	431.3	24.9
±1 SD	1491.4	34.8	88.4	82.3	198.4	1.3
%CV	17.4	60.4	42.4	49.8	46.0	5.4
Sums	188,885.9	1058.8	4477.5	3478.9	9015.2	504.2
Max	12,747.5	153.4	435.0	380.4	922.7	27.1
Min	6599.8	9.6	67.4	45.5	122.5	22.4
Median	9442.2	45.9	203.2	156.9	405.7	25.6

NOTE: Statistics are from 1991 to 2000 only; continuous lake pH data are not available prior to ΔS for H+ is <1% of I-O flux and can be ignored in calculations of the residual term.

Total in	SWO	GWO	Total out	Net balance I-O	Retention =1−(O/I) %	Pool change ΔS	Residual I-O-ΔS
13,485.0	390.9	256.4	647.3	12,837.6	95.2%		12,837.6
10,219.7	253.4	184.9	438.3	9781.4	95.7%		9781.4
10,123.2	100.0	136.1	236.1	9887.1	97.7%		9887.1
10,176.9	168.9	168.4	337.3	9839.6	96.7%		9839.6
9764.4	119.8	206.1	325.9	9438.5	96.7%		9438.5
10,861.7	313.8	268.5	582.3	10,279.4	94.6%		10,279.4
8536.7	115.0	255.0	370.0	8166.7	95.7%		8166.7
9966.5	51.3	231.5	282.8	9683.7	97.2%		9683.7
11,370.7	479.7	403.7	883.4	10,487.3	92.2%		10,487.3
13,503.8	328.9	284.7	613.6	12,890.2	95.5%		12,890.2
10,718.7	135.8	192.2	328.0	10,390.7	96.9%	−87.5	10,478.2
8535.4	168.2	217.9	386.1	8149.3	95.5%	0.2	8149.0
9187.8	159.2	228.1	387.3	8800.6	95.8%	−53.0	8853.5
9794.4	230.9	180.2	411.1	9383.4	95.8%	−34.7	9418.0
9482.1	112.1	178.7	290.8	9191.3	96.9%	68.8	9122.4
10,050.7	252.8	168.1	420.9	9629.8	95.8%	88.4	9541.3
8942.6	224.2	192.5	416.7	8526.0	95.3%	−123.1	8649.1
8100.6	96.4	114.0	210.4	7890.2	97.4%	−16.8	7907.0
6919.7	58.4	97.8	156.2	6763.6	97.7%	−60.6	6824.1
8664.6	113.9	132.1	246.0	8418.5	97.2%	103.7	8314.8
Hydrogen ion budget statistics for 1991 to 2000							
9039.7	155.2	170.1	325.3	8714.3	96.4%	−11.4	8725.8
1075.4	114.7	69.2	95.3	1483.2	0.9%	76.6	1010.1
11.9	73.9	40.7	29.3	n/a	n/a	n/a	n/a
198,405.3	3873.6	4096.9	7970.5	n/a	n/a	n/a	n/a
10,718.7	479.7	403.7	883.4	12,890.2	97.7%	103.7	10,478.2
6919.7	51.3	97.8	156.2	6763.6	92.2%	−123.1	6824.1
9880.5	163.7	192.3	378.1	9534.1	95.8%	−25.7	8751.3

TABLE 3-8G. *Annual sulfate budget, in kilograms/year*

Calendar year	P	NE in	NW in	West in	SWI	GWI
1981	529.4	98.4	2072.9	1280.3	3451.6	650.5
1982	362.3	60.0	1239.1	789.1	2088.2	649.5
1983	389.3	74.2	1514.4	923.1	2511.7	650.8
1984	368.9	93.0	1631.6	959.7	2684.3	681.5
1985	344.9	60.9	1043.4	702.3	1806.6	638.2
1986	393.7	76.6	1553.0	895.0	2524.6	651.6
1987	261.0	50.6	1089.1	652.0	1791.7	653.1
1988	338.6	51.6	806.7	502.9	1361.2	585.9
1989	453.5	86.0	1426.9	835.5	2348.4	590.9
1990	511.2	123.0	2126.5	1102.6	3352.1	673.2
1991	443.4	87.6	1283.7	789.2	2160.5	654.0
1992	344.9	79.6	1267.6	766.5	2113.7	619.9
1993	352.9	71.9	1026.4	630.6	1728.9	606.3
1994	379.4	67.8	1109.3	673.1	1850.2	619.1
1995	326.0	68.1	1124.0	722.5	1914.6	562.5
1996	372.5	127.7	1984.6	1129.5	3241.8	680.5
1997	324.4	79.9	1072.4	670.1	1822.4	666.6
1998	279.6	70.7	1246.8	695.7	2013.2	614.2
1999	265.8	80.9	1051.5	700.8	1833.2	596.5
2000	302.1	80.1	1228.7	762.3	2071.1	627.2
Sulfate budget statistical summary						
Average	367.2	79.4	1344.9	809.1	2233.5	633.6
±1 SD	72.8	20.1	367.6	190.7	572.0	33.4
%CV	19.8	25.3	27.3	23.6	25.6	5.3
Sums	7343.8	1588.6	26,898.6	16,182.8	44,670.0	12,672.1
Max	529.4	127.7	2126.5	1280.3	3451.6	681.5
Min	261.0	50.6	806.7	502.9	1361.2	562.5
Median	357.6	78.1	1242.9	764.4	2079.6	643.9

Total in	SWO	GWO	Total out	Net balance I-O	Retention =1−O/I %	Pool change ΔS	Residual I-O-ΔS
4631.5	2410.4	1668.7	4079.1	552.3	11.9%	−298.1	850.4
3100.0	1198.4	1861.3	3059.7	40.4	1.3%	80.1	−39.7
3551.8	1209.8	1511.4	2721.2	830.6	23.4%	−365.2	1195.8
3734.7	1456.7	1503.7	2960.4	774.3	20.7%	486.7	287.6
2789.6	737.7	1468.8	2206.5	583.1	20.9%	−755.1	1338.2
3570.0	1280.1	1558.2	2838.3	731.6	20.5%	198.1	533.6
2705.7	706.0	1703.6	2409.6	296.1	10.9%	−591.8	887.8
2285.8	418.1	1816.5	2234.6	51.2	2.2%	243.6	−192.4
3392.8	1109.0	1627.2	2736.2	656.6	19.4%	−311.4	968.0
4536.5	1579.7	1400.9	2980.6	1555.9	34.3%	593.6	962.2
3257.9	1074.8	1740.5	2815.3	442.6	13.6%	−303.5	746.1
3078.5	1125.1	1634.1	2759.2	319.3	10.4%	2.7	316.6
2688.1	1010.1	1786.1	2796.2	−108.1	−4.0%	−187.0	78.9
2848.7	984.0	1594.5	2578.5	270.2	9.5%	−122.9	393.1
2803.1	823.9	1361.5	2185.4	617.6	22.0%	108.4	509.2
4294.8	2081.5	1374.1	3455.6	839.2	19.5%	−134.8	974.0
2813.4	1166.4	1715.5	2881.9	−68.5	−2.4%	−117.3	48.8
2907.0	985.7	1376.9	2362.6	544.4	18.7%	−79.2	623.6
2695.4	826.6	1534.4	2361.0	334.5	12.4%	7.5	327.0
3000.4	1070.5	1440.0	2510.5	489.9	16.3%	43.8	446.1
Sulfate budget statistical summary							
3234.3	1162.7	1583.9	2746.6	487.7	14.1%	−75.1	562.8
646.6	456.4	155.3	451.5	382.4	9.5%	325.7	422.4
20.0	39.2	9.8	16.4	n/a	n/a	n/a	n/a
64,685.9	23,254.5	31,678.2	54,932.6	n/a	n/a	n/a	n/a
4631.5	2410.4	1861.3	4079.1	1555.9	34.3%	593.6	1338.2
2285.8	418.1	1361.5	2185.4	−108.1	−4.0%	−755.1	−192.4
3039.5	1091.9	1576.3	2747.7	517.1	15.0%	−98.2	521.4

TABLE 3-8H. *Annual nitrate budget, in kilograms/year*

Calendar year	P	NE in	NW in	West in	SWI	GWI
1981	325.4	2.8	42.8	35.9	81.5	0.9
1982	292.3	1.2	18.8	26.5	46.5	0.9
1983	244.3	0.9	9.3	28.8	39.0	0.9
1984	268.5	0.7	9.0	27.1	36.8	1.0
1985	282.4	0.6	6.5	27.5	34.7	0.9
1986	286.4	0.7	10.1	26.4	37.3	0.9
1987	232.1	0.7	9.8	34.7	45.2	0.9
1988	295.9	0.7	13.4	30.7	44.7	0.8
1989	357.1	0.4	23.0	26.2	49.7	0.8
1990	402.6	0.7	6.1	13.9	20.6	1.0
1991	328.0	1.0	14.1	26.7	41.8	0.9
1992	300.8	1.9	9.1	21.5	32.5	0.9
1993	299.9	0.3	2.6	7.7	10.7	0.9
1994	313.7	0.3	2.6	7.0	10.0	0.9
1995	330.5	0.3	2.2	4.4	6.9	0.8
1996	336.4	0.3	5.7	5.0	11.0	1.0
1997	320.0	0.2	3.5	4.6	8.3	0.9
1998	279.4	0.3	4.0	5.3	9.6	0.9
1999	242.2	0.5	10.9	5.2	16.7	0.8
2000	252.2	0.8	7.4	8.1	16.3	0.9
Nitrate budget statistical summary						
Average	299.5	0.8	10.6	18.7	30.0	0.9
±1 SD	41.9	0.6	9.3	11.6	19.3	0.1
%CV	14.0	81.1	88.2	61.9	64.5	5.3
Sums	5990.2	15.3	211.0	373.3	599.6	18.0
Max	402.6	2.8	42.8	35.9	81.5	1.0
Min	232.1	0.2	2.2	4.4	6.9	0.8
Median	297.9	0.7	9.0	23.9	33.6	0.9

Total in	SWO	GWO	Total out	Net balance I-O	Retention =1−(O/I) %	Pool change ΔS	Residual I-O-ΔS
407.8	40.7	23.0	63.8	344.0	84.4%	−38.8	382.8
339.8	72.0	36.7	108.8	231.0	68.0%	13.0	218.0
284.3	18.7	17.4	36.1	248.2	87.3%	4.4	243.8
306.2	37.7	27.3	65.0	241.2	78.8%	0.1	241.2
318.0	18.6	15.7	34.2	283.8	89.2%	38.6	245.2
324.5	45.0	32.4	77.4	247.1	76.1%	−38.8	285.9
278.3	18.7	23.7	42.4	235.8	84.7%	25.9	209.9
341.4	17.3	44.6	61.8	279.6	81.9%	4.2	275.4
407.6	77.5	63.6	141.0	266.5	65.4%	43.3	223.2
424.2	44.5	40.4	84.9	339.3	80.0%	−38.5	377.8
370.7	17.5	23.2	40.8	330.0	89.0%	−30.5	360.5
334.2	26.2	26.3	52.5	281.7	84.3%	−17.3	299.1
311.4	3.9	6.9	10.8	300.6	96.5%	4.2	296.4
324.6	17.1	25.2	42.3	282.3	87.0%	−8.7	291.0
338.2	3.5	6.9	10.4	327.8	96.9%	4.3	323.5
348.4	21.0	17.4	38.4	309.9	89.0%	8.8	301.1
329.3	61.1	43.2	104.3	225.0	68.3%	−0.2	225.2
289.9	9.6	9.4	18.9	271.0	93.5%	30.3	240.6
259.7	15.5	22.7	38.2	221.6	85.3%	−43.2	264.8
269.4	10.0	13.9	23.9	245.5	91.1%	51.9	193.6
Nitrate budget statistical summary							
330.4	28.8	26.0	54.8	275.6	83.8%	0.6	274.9
45.5	21.7	14.2	34.4	39.2	8.9%	28.7	55.0
13.8	75.4	54.8	62.7	n/a	n/a	n/a	n/a
6607.8	575.9	520.1	1095.9	n/a	n/a	n/a	n/a
424.2	77.5	63.6	141.0	344.0	96.9%	51.9	382.8
259.7	3.5	6.9	10.4	221.6	65.4%	−43.2	193.6
327.0	18.7	23.5	42.4	275.3	85.0%	4.2	270.1

TABLE 3-81. *Annual chloride budget, in kilograms/year*

Calendar year	P	NE in	NW in	West in	SWI	GWI
1981	55.2	608.5	300.9	394.5	1303.9	127.1
1982	27.6	311.2	144.2	201.9	657.3	126.9
1983	33.7	454.1	207.2	277.1	938.4	127.2
1984	33.3	509.8	223.2	314.8	1047.8	133.2
1985	28.9	370.1	154.8	220.4	745.3	124.7
1986	29.6	452.2	209.1	331.5	992.8	127.3
1987	35.4	344.8	203.6	244.7	793.1	127.6
1988	35.5	402.2	141.5	188.3	732.0	114.5
1989	42.4	670.8	294.2	351.8	1316.8	115.5
1990	40.3	915.0	450.7	608.3	1974.0	131.5
1991	26.4	743.3	280.9	383.8	1408.0	127.8
1992	40.3	720.1	257.7	372.0	1349.8	121.1
1993	58.6	692.5	223.8	274.4	1190.7	118.5
1994	27.0	634.0	257.9	333.6	1225.5	121.0
1995	52.5	635.8	294.1	382.5	1312.4	109.9
1996	85.1	1004.6	517.5	641.5	2163.6	133.0
1997	33.8	610.2	267.2	367.6	1245.0	130.2
1998	24.3	656.6	315.2	401.5	1373.3	120.0
1999	46.8	624.4	330.4	532.8	1487.6	116.5
2000	32.9	563.7	459.2	692.4	1715.3	122.5
Chloride budget statistical summary						
Average	39.5	596.2	276.7	375.8	1248.6	123.8
±1 SD	14.5	179.3	102.4	142.4	396.3	6.5
%CV	36.8	30.1	37.0	37.9	31.7	5.3
Sums	789.7	11,923.9	5533.3	7515.4	24,972.6	2475.9
Max	85.1	1004.6	517.5	692.4	2163.6	133.2
Min	24.3	311.2	141.5	188.3	657.3	109.9
Median	34.6	617.3	262.5	359.7	1274.5	125.8

Total in	SWO	GWO	Total out	Net balance I-O	Retention =1−(O/I) %	Pool change ΔS	Residual I-O-ΔS
1486.2	981.0	672.3	1653.3	−167.2	−11.2%	−141.6	−25.6
811.8	377.9	733.1	1111.0	−299.2	−36.9%	83.9	−383.1
1099.3	532.0	689.6	1221.6	−122.3	−11.1%	63.8	−186.1
1214.3	639.9	697.4	1337.3	−123.1	−10.1%	282.1	−405.2
898.9	370.8	754.0	1124.8	−225.9	−25.1%	−49.1	−176.8
1149.7	568.0	769.1	1337.1	−187.4	−16.3%	4.4	−191.8
956.1	391.3	958.8	1350.1	−394.0	−41.2%	14.9	−408.9
882.0	243.7	1089.7	1333.4	−451.4	−51.2%	116.7	−568.1
1474.7	650.9	1005.5	1656.4	−181.7	−12.3%	31.0	−212.7
2145.9	1008.3	888.6	1896.9	249.0	11.6%	361.2	−112.2
1562.2	725.6	1170.0	1895.6	−333.4	−21.3%	−66.5	−266.8
1511.3	751.3	1103.9	1855.2	−344.0	−22.8%	288.2	−632.2
1367.8	685.0	1299.8	1984.8	−617.0	−45.1%	−149.4	−467.7
1373.5	662.1	1244.8	1906.9	−533.4	−38.8%	110.8	−644.2
1474.8	695.2	1146.9	1842.1	−367.3	−24.9%	57.3	−424.6
2381.6	1570.1	1029.5	2599.6	−218.0	−9.2%	−268.9	50.9
1409.0	735.7	1254.8	1990.5	−581.5	−41.3%	241.4	−822.9
1517.6	738.4	1087.0	1825.4	−307.9	−20.3%	−132.0	−175.9
1650.9	639.8	1234.2	1874.0	−223.1	−13.5%	137.2	−360.4
1870.8	914.6	1261.6	2176.2	−305.5	−16.3%	130.2	−435.6
		Chloride budget statistical summary					
1411.9	694.1	1004.5	1698.6	−286.7	−22.9%	55.8	−342.5
405.5	286.1	218.8	386.8	191.5	15.5%	163.1	221.4
28.7	41.2	21.8	22.8	n/a	n/a	n/a	n/a
28,238.2	13,881.6	20,090.7	33,972.3	n/a	n/a	n/a	n/a
2381.6	1570.1	1299.8	2599.6	249.0	11.6%	361.2	50.9
811.8	243.7	672.3	1111.0	−617.0	−51.2%	−268.9	−822.9
1441.8	673.5	1058.3	1833.8	−302.3	−20.8%	60.5	−371.7

TABLE 3-8J. *Annual phosphate budget, in kilograms/year*

Calendar year	P	NE in	NW in	West in	SWI	GWI
1981	3.5	0.3	1.5	2.5	4.2	0.3
1982	2.1	0.2	1.4	2.0	3.6	0.3
1983	2.5	0.2	1.1	1.6	2.9	0.3
1984	3.0	0.2	1.6	1.8	3.7	0.4
1985	2.7	0.1	1.0	1.1	2.2	0.3
1986	3.7	0.2	2.3	2.3	4.8	0.3
1987	4.6	0.2	2.0	3.2	5.4	0.3
1988	5.0	0.2	1.8	2.2	4.2	0.3
1989	1.9	0.1	1.1	1.1	2.2	0.3
1990	4.1	0.3	3.3	2.9	6.5	0.4
1991	6.4	0.1	1.1	1.2	2.4	0.3
1992	5.6	0.1	0.9	1.1	2.0	0.3
1993	1.8	0.0	0.4	0.4	0.8	0.3
1994	2.0	0.1	0.4	0.6	1.0	0.3
1995	8.2	0.1	0.5	0.6	1.1	0.3
1996	1.9	0.1	0.7	0.9	1.7	0.4
1997	2.4	0.1	0.4	0.4	0.9	0.4
1998	3.5	0.1	0.9	0.7	1.6	0.3
1999	6.8	0.1	0.7	0.8	1.6	0.3
2000	2.0	0.1	0.5	0.6	1.2	0.3
Phosphate budget statistical summary						
Average	3.7	0.1	1.2	1.4	2.7	0.3
±1 SD	1.9	0.1	0.7	0.9	1.6	0.0
%CV	50.8	52.9	64.0	62.5	60.9	5.3
Sums	73.8	2.8	23.4	28.0	54.2	6.8
Max	8.2	0.3	3.3	3.2	6.5	0.4
Min	1.8	0.0	0.4	0.4	0.8	0.3
Median	3.2	0.1	1.0	1.1	2.2	0.3

Total in	SWO	GWO	Total out	Net balance I-O	Retention =1−(O/I) %	Pool change ΔS	Residual I-O-ΔS
8.1	4.8	3.2	7.9	0.1	1.5%	1.8	−1.7
6.0	1.8	2.8	4.6	1.4	24.0%	−1.2	2.7
5.8	1.9	2.4	4.4	1.4	24.2%	−0.7	2.1
7.1	2.6	2.8	5.4	1.7	23.7%	−7.1	8.8
5.3	1.3	2.9	4.2	1.0	19.2%	11.4	−10.4
8.9	2.4	3.5	5.9	3.0	34.0%	−0.7	3.7
10.4	2.1	5.6	7.8	2.6	24.9%	12.6	−10.0
9.6	2.1	5.8	7.9	1.7	17.5%	−21.9	23.5
4.5	1.4	2.0	3.4	1.1	24.9%	6.5	−5.4
11.0	3.7	3.2	6.9	4.1	37.1%	−1.5	5.6
9.2	1.7	2.8	4.5	4.7	51.4%	−2.8	7.5
8.0	1.6	2.3	3.9	4.1	50.9%	−0.4	4.5
2.9	0.8	1.4	2.3	0.6	22.0%	−2.0	2.6
3.4	0.6	1.1	1.8	1.6	47.6%	0.3	1.4
9.6	0.7	1.0	1.7	7.9	82.1%	0.9	7.1
4.0	1.2	0.8	1.9	2.0	51.2%	−2.2	4.2
3.6	0.4	0.8	1.2	2.4	67.6%	−0.0	2.4
5.5	0.7	1.1	1.8	3.7	67.4%	2.6	1.1
8.7	0.8	1.5	2.3	6.4	73.4%	−1.7	8.1
3.5	0.7	1.0	1.7	1.7	50.5%	2.8	−1.0
Phosphate budget statistical summary							
6.7	1.7	2.4	4.1	2.7	39.7%	−0.2	2.8
2.6	1.1	1.4	2.3	2.0	21.7%	6.9	7.2
38.8	65.8	59.8	55.9	n/a	n/a	n/a	n/a
134.8	33.4	48.1	81.5	n/a	n/a	n/a	n/a
11.0	4.8	5.8	7.9	7.9	82.1%	12.6	23.5
2.9	0.4	0.8	1.2	0.1	1.5%	−21.9	−10.4
6.5	1.5	2.4	4.1	1.9	35.5%	−0.5	2.7

TABLE 3-8K. *Annual dissolved organic carbon (DOC) budget, in kilograms/year*

Calendar year	P	NE in	NW in	West in	SWI	GWI
1981	214.0	n/a	n/a	n/a		101.8
1982	144.5	n/a	n/a	n/a		101.6
1983	188.2	n/a	n/a	n/a		101.8
1984	178.0	n/a	n/a	n/a		106.6
1985	148.3	n/a	n/a	n/a		99.9
1986	178.9	n/a	n/a	n/a		102.0
1987	155.6	n/a	n/a	n/a		102.2
1988	130.0	n/a	n/a	n/a		91.7
1989	183.0	n/a	n/a	n/a		92.5
1990	219.7	n/a	n/a	n/a		105.3
1991	172.6	n/a	n/a	n/a		102.3
1992	163.1	n/a	134.4	n/a		97.0
1993	149.0	58.4	285.8	323.0	667.2	94.9
1994	153.7	69.6	352.1	398.3	820.0	96.9
1995	170.7	68.7	520.1	n/a	n/a	88.0
1996	225.1	n/a	n/a	n/a	n/a	106.5
1997	157.5	n/a	n/a	n/a	n/a	104.3
1998	177.9	114.8	487.7	622.9	1225.4	96.1
1999	162.5	101.3	333.1	419.0	853.4	93.3
2000	168.5	116.2	382.8	526.0	1025.0	98.2
DOC budget statistical summary						
Average	172.0	88.2	356.6	457.8	918.2	99.1
±1 SD	25.1	25.6	128.8	117.4	213.7	5.2
%CV	14.6	29.0	36.1	25.6	23.3	5.3
Sums	3440.9	529.0	2496.0	2289.2	4591.0	1983.0
Max	225.1	116.2	520.1	622.9	1225.4	106.6
Min	130.0	58.4	134.4	323.0	667.2	88.0
Median	169.6	85.4	352.1	419.0	853.4	100.8

NOTE: DOC data not available prior to 1992 and incomplete for 1995 to 1997.

Total in	SWO	GWO	Total out	Net balance I-O	Retention =1−(O/I) %	Pool change ΔS	Residual I-O-ΔS
n/a	n/a	n/a	n/a	n/a		n/a	n/a
n/a	n/a	n/a	n/a	n/a		n/a	n/a
n/a	n/a	n/a	n/a	n/a		n/a	n/a
n/a	n/a	n/a	n/a	n/a		n/a	n/a
n/a	n/a	n/a	n/a	n/a		n/a	n/a
n/a	n/a	n/a	n/a	n/a		n/a	n/a
n/a	n/a	n/a	n/a	n/a		n/a	n/a
n/a	n/a	n/a	n/a	n/a		n/a	n/a
n/a	n/a	n/a	n/a	n/a		n/a	n/a
n/a	n/a	n/a	n/a	n/a		n/a	n/a
n/a	n/a	n/a	n/a	n/a		n/a	n/a
n/a	n/a	n/a	n/a	n/a		n/a	n/a
911.1	n/a	n/a	n/a	n/a		n/a	n/a
1070.6	434.2	654.2	1088.4	−17.9	−1.7%	0.0	−17.9
n/a	584.4	910.1	1494.5	n/a	n/a	n/a	n/a
n/a	772.9	477.2	1250.1	n/a	n/a	n/a	n/a
n/a	560.6	364.4	925.0	n/a	n/a	n/a	n/a
1499.5	493.7	733.0	1226.7	272.7	18.2%	0.0	272.7
1109.2	418.2	808.9	1227.1	−117.8	−10.6%	0.0	−117.8
1291.6	489.9	748.0	1237.9	53.7	4.2%	0.0	53.7
DOC budget statistical summary							
1176.4	536.3	670.8	1207.1	47.7	2.5%	0.0	47.7
225.7	120.6	190.5	173.1	165.7	12.1%	0.0	165.7
19.2	22.5	28.4	14.3	n/a	n/a	n/a	n/a
5882.0	3753.9	4695.7	8449.6	n/a	n/a	n/a	n/a
1499.5	772.9	910.1	1494.5	272.7	18.2%	0.0	272.7
911.1	418.2	364.4	925.0	−117.8	−10.6%	0.0	−117.8
1109.2	493.7	733.0	1227.1	17.9	1.2%	0.0	17.9

TABLE 3-8L. *Annual dissolved inorganic carbon (DIC) budget, in mol/year*

Calendar year	P	NE in	NW in	West in	Net Retention SWI	Pool GWI
1981	7884.0	n/a	n/a	n/a	n/a	82,613.6
1982	5325.0	n/a	n/a	n/a	n/a	82,490.6
1983	6935.0	n/a	n/a	n/a	n/a	82,653.9
1984	6557.0	n/a	n/a	n/a	n/a	86,550.1
1985	5463.0	n/a	n/a	n/a	n/a	81,045.5
1986	6591.0	n/a	n/a	n/a	n/a	82,754.8
1987	5731.0	n/a	n/a	n/a	n/a	82,936.4
1988	4790.0	n/a	n/a	n/a	n/a	74,413.6
1989	6743.0	n/a	n/a	n/a	n/a	75,044.1
1990	8095.0	n/a	n/a	n/a	n/a	85,496.5
1991	6360.0	n/a	n/a	n/a	n/a	83,054.2
1992	6008.0	n/a	n/a	n/a	n/a	78,732.5
1993	5489.0	n/a	n/a	n/a	n/a	76,996.8
1994	5662.0	3472.0	22,051.0	20,660.0	46,183.0	78,619.1
1995	6290.0	3341.0	19,403.0	17,342.0	40,086.0	71,432.5
1996	8292.0	6338.0	31,505.0	32,148.0	69,991.0	86,426.8
1997	5803.0	4710.0	23,204.0	22,809.0	50,723.0	84,660.3
1998	6556.0	5021.0	24,841.0	22,169.0	52,031.0	78,003.6
1999	5987.0	3802.0	20,824.0	19,602.0	44,228.0	75,750.6
2000	6207.0	3620.0	20,834.0	20,710.0	45,164.0	79,657.4
DIC budget statistical summary						
Average	6399.6	4329.1	23,237.4	22,205.7	49,772.3	79,221.4
±1 SD	924.7	1091.7	4054.6	4730.7	9776.5	4244.8
%CV	14.5	25.2	17.4	21.3	19.6	5.4
Sums	126,768.0	30,304.0	162,662.0	155,440.0	348,406.0	1,609,332.7

NOTE: Routine DIC data not available prior to 1994; lake DIC available in 1992.

Total in	SWO	GWO	Total out	balance I-O	=1−(O/I) %	change ΔS	Residual I-O-ΔS
n/a	n/a	n/a	n/a	n/a		n/a	n/a
n/a	n/a	n/a	n/a	n/a		n/a	n/a
n/a	n/a	n/a	n/a	n/a		n/a	n/a
n/a	n/a	n/a	n/a	n/a		n/a	n/a
n/a	n/a	n/a	n/a	n/a		n/a	n/a
n/a	n/a	n/a	n/a	n/a		n/a	n/a
n/a	n/a	n/a	n/a	n/a		n/a	n/a
n/a	n/a	n/a	n/a	n/a		n/a	n/a
n/a	n/a	n/a	n/a	n/a		n/a	n/a
n/a	n/a	n/a	n/a	n/a		n/a	n/a
n/a	n/a	n/a	n/a	n/a		n/a	n/a
n/a	n/a	n/a	n/a	n/a		8748	n/a
n/a	n/a	n/a	n/a	n/a		3912	n/a
130,464.1	34,178.0	52,146.4	86,324.4	44,139.7	33.8%	12,899	31,241.2
117,808.5	26,544.0	45,777.4	72,321.4	45,487.1	38.6%	4180	41,306.9
164,709.8	59,926.0	41,051.8	100,977.8	63,732.0	38.7%	-3573	67,304.8
141,186.3	43,004.0	59,399.1	102,403.1	38,783.1	27.5%	-5123	43,906.1
136,590.6	29,752.0	43,504.5	73,256.5	63,334.0	46.4%	-8383	71,717.0
125,965.6	28,363.0	50,825.1	79,188.1	46,777.5	37.1%	-11274	58,051.3
131,028.4	36,802.0	48,347.1	85,149.1	45,879.2	35.0%	28539	17,340.7

DIC budget statistical summary

Total in	SWO	GWO	Total out	balance I-O	=1−(O/I) %	change ΔS	Residual I-O-ΔS
135,393.3	36,938.4	487,21.6	85,660.1	49,733.2	36.7%	2466.4	47,266.9
14,921.5	11,590.8	6124.0	12,174.6	9778.9	5.7%	14,095.5	19,642.0
11.0	31.4	12.6	14.2	n/a	n/a	n/a	n/a
947,753.1	258,569.0	341,051.4	599,620.4	n/a	n/a	n/a	n/a

TABLE 3-8M. *Annual acid-neutralizing capacity (ANC) budget, in eq/year*

Calendar year	P	NE in	NW in	West in	SWI	GWI
1981	−12,536.4	n/a	n/a	n/a	n/a	36,973.9
1982	−9747.3	n/a	n/a	n/a	n/a	36,918.9
1983	−9631.9	n/a	n/a	n/a	n/a	36,992.0
1984	−9719.6	n/a	n/a	n/a	n/a	38,735.7
1985	−9488.6	n/a	n/a	n/a	n/a	36,272.1
1986	−10,280.4	n/a	n/a	n/a	n/a	37,037.1
1987	−8219.2	n/a	n/a	n/a	n/a	37,118.4
1988	−9820.7	n/a	n/a	n/a	n/a	33,304.0
1989	−10,859	n/a	n/a	n/a	n/a	33,586.2
1990	−12,747.5	n/a	n/a	n/a	n/a	38,264.2
1991	−10,376.8	515.0	10,784.0	7508.0	18,807.0	37,171.1
1992	−8142.6	397.0	11,681.0	11,036.0	23,114.0	35,236.9
1993	−8787.3	620.0	10,289.0	8210.0	19,119.0	34,460.1
1994	−9395.9	671.0	11,672.0	10,293.0	22,636.0	35,186.2
1995	−9134.9	737.0	9261.0	7676.0	17,674.0	31,969.8
1996	−9173.2	624.0	11,416.0	11,568.0	23,608.0	38,680.5
1997	−8511.7	717.0	10,598.0	9235.0	20,550.0	37,889.9
1998	−7669.2	1035.0	12,900.0	10,797.0	24,732.0	34,910.7
1999	−6599.8	665.0	11,250.0	9471.0	21,386.0	33,902.4
2000	−8043.9	603.0	9461.0	8101.0	18,165.0	35,650.8
ANC budget statistical summary						
Average	−8384.3	674.3	10,947.6	9598.6	21,220.4	35,320.8
±1 SD	1491.4	165.4	1096.5	1483.7	2487.1	1899.8
%CV	n/a	24.5	10.0	15.5	11.7	5.4
Sums	−188,885.9	6584.0	109,312.0	93,895.0	209,791.0	720,260.8
Max	−6599.8	1035.0	12,900.0	11,568.0	24,732.0	38,735.7
Min	−12,747.5	397.0	9261.0	7508.0	17,674.0	31,969.8
Median	−9442.2	644.5	11,017.0	9353.0	20,968.0	36,595.5

NOTE: Routine ANC data not available prior to 1991.

Total in	SWO	GWO	Total out	Net balance I-O	Retention =1−(O/I) %	Pool change ΔS	Residual I-O-ΔS
n/a	n/a	n/a	n/a	n/a		n/a	n/a
n/a	n/a	n/a	n/a	n/a		n/a	n/a
n/a	n/a	n/a	n/a	n/a		n/a	n/a
n/a	n/a	n/a	n/a	n/a		n/a	n/a
n/a	n/a	n/a	n/a	n/a		n/a	n/a
n/a	n/a	n/a	n/a	n/a		n/a	n/a
n/a	n/a	n/a	n/a	n/a		n/a	n/a
n/a	n/a	n/a	n/a	n/a		n/a	n/a
n/a	n/a	n/a	n/a	n/a		n/a	n/a
n/a	n/a	n/a	n/a	n/a		n/a	n/a
45,601.3	n/a	16,371.3	n/a	n/a		5934	n/a
50,208.3	17,214.0	25,382.1	42,596.1	7612.2	15.2%	−3421	11,033.2
44,791.8	16,073.0	30,123.0	46,196.0	−1404.2	−3.1%	9263	−10,666.9
48,426.3	15,334.0	29,871.5	45,205.5	3220.7	6.7%	2552	668.9
40,508.9	15,511.0	25,521.5	41,032.5	−523.7	−1.3%	−7850	7326.6
53115.3	27,496.0	17,519.2	45,015.2	8100.1	15.3%	−17,817	25,917.2
49,928.2	15,410.0	26,556.6	41,966.6	7961.6	15.9%	19,505	−11,543.4
51,973.5	17,006.0	26,057.5	43,063.5	8910.0	17.1%	1017	7892.6
48,688.6	16,154.0	31,287.4	47,441.4	1247.1	2.6%	4295	−3047.6
45,771.9	20,296.0	28,633.2	48,929.2	−3157.3	−6.9%	−4308	1150.3

ANC budget statistical summary

48,157.0	17,832.7	26,772.5	44,605.1	3551.8	6.8%	359.5	3192.3
3774.7	3935.9	5074.1	2644.3	4702.9	9.4%	10,609.0	11,596.2
7.8	22.1	19.0	5.9	n/a	n/a	n/a	n/a
479,014.1	160,494.0	257,323.5	401,446.2	n/a	n/a	n/a	n/a
53,115.3	27,496.0	31,287.4	48,929.2	8910.0	17.1%	19,505.0	25,917.2
40,508.9	15,334.0	16,371.3	41,032.5	−3157.3	−6.9%	−17,817.2	−11,543.4
48,557.4	16,154.0	26,307.0	45,015.2	3220.7	6.7%	1784.6	1150.3

TABLE 3-8N. *Dissolved silica (silicate, SiO_2) budget, in kilograms/year*

Calendar year	P	NE in	NW in	West in	SWI	GWI
1981	11.8	198.7	2400.0	1894.7	4493.4	1294.1
1982	8.6	113.4	1539.5	1221.4	2874.3	1292.2
1983	10.6	124.9	1565.0	1137.0	2826.9	1294.7
1984	11.2	162.1	1875.3	1357.8	3395.2	1355.7
1985	9.1	98.8	1193.3	951.5	2243.6	1269.5
1986	15.2	129.1	1746.6	1339.3	3215.0	1296.3
1987	10.9	93.3	1353.3	1075.2	2521.8	1299.1
1988	11.2	79.0	1019.4	746.6	1845.0	1165.6
1989	13.5	128.6	1567.4	1116.3	2812.3	1175.5
1990	14.4	206.2	2355.6	1606.9	4168.7	1339.2
1991	10.6	146.5	1578.7	1192.1	2917.3	1301.0
1992	9.6	129.9	1470.0	1180.9	2780.8	1233.3
1993	8.5	109.9	1201.8	899.8	2211.5	1206.1
1994	10.7	108.4	1336.7	1029.5	2474.6	1231.5
1995	11.9	105.4	1417.8	1097.1	2620.3	1118.9
1996	17.7	203.2	2317.2	1692.6	4213.0	1353.8
1997	9.0	138.5	1381.9	1068.9	2589.3	1326.1
1998	10.1	146.8	1732.5	1274.3	3153.6	1221.9
1999	10.2	131.9	1413.3	1195.3	2740.5	1186.6
2000	9.3	128.5	1646.2	1236.9	3011.6	1247.8
SIO_2 budget statistical summary						
Average	11.2	134.2	1605.6	1215.7	2955.4	1260.5
±1 SD	2.4	35.5	382.0	269.4	680.7	66.5
%CV	21.2	26.4	23.8	22.2	23.0	5.3
Sums	224.1	2683.1	32,111.5	24,314.1	59,108.7	25,209.1
Max	17.7	206.2	2400.0	1894.7	4493.4	1355.7
Min	8.5	79.0	1019.4	746.6	1845.0	1118.9
Median	10.6	128.8	1552.2	1186.5	2819.6	1280.8

Total in	SWO	GWO	Total out	Net balance I-O	Retention =1−(O/I) %	Pool change ΔS	Residual I-O-ΔS
5799.3	883.0	605.0	1488.0	4311.3	74.3%	618.2	3693.1
4175.0	784.9	928.1	1713.0	2462.0	59.0%	464.8	1997.3
4132.3	538.6	603.1	1141.7	2990.5	72.4%	−152.5	3143.1
4762.2	799.0	673.8	1472.8	3289.4	69.1%	−230.2	3519.6
3522.3	299.8	509.8	809.6	2712.7	77.0%	−140.8	2853.4
4526.5	740.5	788.0	1528.5	2998.0	66.2%	824.5	2173.5
3831.8	428.8	941.9	1370.7	2461.1	64.2%	−434.4	2895.5
3021.9	207.8	728.6	936.4	2085.5	69.0%	−658.8	2744.3
4001.3	508.7	643.0	1151.7	2849.6	71.2%	296.7	2552.9
5522.3	845.3	765.8	1611.1	3911.2	70.8%	808.8	3102.4
4228.9	633.5	997.0	1630.5	2598.3	61.4%	−178.6	2777.0
4023.7	647.8	909.3	1557.1	2466.6	61.3%	−302.0	2768.6
3426.1	583.8	868.6	1452.4	1973.7	57.6%	−603.8	2577.5
3716.8	513.6	653.5	1167.1	2549.7	68.6%	457.1	2092.6
3751.1	313.6	510.1	823.7	2927.4	78.0%	−166.0	3093.4
5584.5	1215.4	774.3	1989.7	3594.9	64.4%	409.2	3185.7
3924.5	815.2	1044.6	1859.8	2064.7	52.6%	−315.2	2379.9
4385.6	578.2	765.6	1343.8	3041.8	69.4%	124.8	2917.0
3937.3	471.0	767.8	1238.8	2698.5	68.5%	−389.5	3088.0
4268.6	615.5	691.1	1306.6	2962.0	69.4%	475.6	2486.4
			SIO_2 budget statistical summary				
4227.1	621.2	758.5	1379.7	2847.4	67.2%	45.4	2802.1
722.2	234.6	153.1	317.9	594.7	6.5%	463.5	446.9
17.1	37.8	20.2	23.0	n/a	n/a	n/a	n/a
84,542.0	12,424.0	15,169.0	27,593.0	n/a	n/a	n/a	n/a
5799.3	1215.4	1044.6	1989.7	4311.3	78.0%	824.5	3693.1
3021.9	207.8	509.8	809.6	1973.7	52.6%	−658.8	1997.3
4078.0	599.6	765.7	1411.6	2781.1	68.8%	−146.7	2815.2

TABLE 3-9. *Summary of average annual fluxes, budgets, and residuals, 1981 to 2000*

Flux	Flow	Ca	Mg	K	Na	NH4	H
Units	m³ yr⁻¹	kg yr⁻¹	kg yr⁻¹	kg yr⁻¹	kg yr⁻¹	kg yr⁻¹	eq yr⁻¹
Years	20	20	20	20	20	20	10
Annual average flux							
Inputs							
GWI	112,522	442.3	88.9	139.0	325.8	2.25	24.9
NE in	12,140	100.0	21.4	15.5	282.1	0.5	57.6
NW in	248,049	498.2	103.4	71.5	329.6	3.65	208.5
West in	157,271	355.2	77.2	73.1	326.8	3.28	165.2
P	181,098	14.4	4.4	8.6	19.3	42.8	8583.5
Total inputs	711,080	1410.2	295.3	307.6	1283.6	52.52	9039.7
Outputs							
GWO	348,896	801.1	173.9	164.2	741.7	7.92	170.1
SWO	257,512	564.9	121.3	113.2	514.8	6.31	155.2
E	73856						
Total outputs	680,265	1365.9	295.2	277.4	1256.5	14.24	325.3
Average in lake	865,068	2086.8	453.3	411.1	1921.6	20.0	139.1
Budget							
Imbalance = I-O	30,816	44.2	0.1	30.2	27.1	38.29	8714.4
Change in lake ΔS	−15	−22.9	−4.3	−4.7	31.1	−1.38	−11.4
Residual = (I-O-ΔS)	30,831	67.1	4.4	35.0	−4.1	39.7	8725.8
Residual as % lake	3.6	3.2	1.0	8.5	−0.2	198.3	6273.0
Residual as % inputs	4.3	4.8	1.5	11.4	−0.3	75.5	96.5
Retention = 1−(O/I)	4.3%	3.1%	0.0%	9.8%	2.1%	72.9%	96.4%
Interpretation							
Relative flux	small	small	tiny	moderate	tiny	large	huge
Gain / loss	gain	gain	loss	gain	gain	gain	gain
Sink / source	sink	sink	source	sink	sink	sink	sink
Lake process?	~level ±pool	sediment burial	balanced	biological uptake	balanced	biological uptake	buffered by ANC

NOTE: Routine lake profile DOC values not available to calculate lake pool changes, value set = 0.0 (no change). Lake standing pool DOC estimated to equal average lake outlet DOC × lake volume. DIC and ANC values derived from 1991 to 2000 measured data (10 years).

	SO4	NO3	Cl	PO4	DOC	DIC	ANC	Si
	kg yr^{-1}	kg yr^{-1}	kg yr^{-1}	kg yr^{-1}	kg yr^{-1}	mol yr^{-1}	eq yr^{-1}	kg yr^{-1}
	20	20	20	20	20	7	9	20
Annual average flux								
	633.6	0.90	123.8	0.34	99.1	79,221	35321	1260.5
	79.4	0.8	596.2	0.1	88.2	4329	674	134.2
	1344.9	10.55	276.7	1.17	393.6	23,237	10948	1605.6
	809.1	18.66	375.8	1.40	457.8	22,206	9599	1215.7
	367.2	299.5	39.5	3.7	172.0	6400	−8384	11.2
	3234.3	330.39	1411.9	6.74	1210.8	135,393	48157	4227.1
	1583.9	26.00	1004.5	2.40	670.8	48,722	26773	758.5
	1162.7	28.79	694.1	1.67	536.3	36,938	17833	621.2
	2746.6	54.80	1698.6	4.07	1207.1	85,660	44,605	1379.7
	4029.1	41.7	2605.3	5.90	1795.9	119,337	71,666	1909.8
Budget								
	487.7	275.59	−286.7	2.67	3.7	49,733	3552	2847.4
	−31.7	0.65	55.8	−0.17	0.0	2466	360	45.4
	519.4	274.9	−342.5	2.8	3.7	47,266.8	3192	2802.1
	12.9	659.3	−13.1	48.0	0.2	39.6	4.5	146.7
	16.1	83.2	−24.3	42.1	0.3	34.9	6.6	66.3
	15.1%	83.4%	−20.3%	39.6%	0.3%	36.7%	7.4%	67.4%
Interpretation								
	moderate gain sink biological reduction?	huge gain sink biological uptake	moderate loss source sediment release?	very large gain sink biological uptake	tiny gain sink balanced	very large gain sink biological uptake	small gain sink neutralize H-ion	huge gain sink sediment burial

TABLE 3-10. *Average, standard deviation, minimum, and maximum values of mass in the lake, difference in mass between input and output, and change in mass in the lake, 1981 to 2000*

Chemical	Number of years	Average	Standard deviation	Minimum	Maximum
Calcium					
Lake	20	2088.70	94.30	1929.62	2318.60
I-O	20	44.22	151.82	−209.50	409.50
dM	20	−22.86	126.37	−287.70	210.40
Magnesium					
Lake	20	455.88	16.92	423.45	493.13
I-O	20	0.08	29.95	−52.00	72.60
dM	20	−4.32	25.05	−59.50	49.20
Potassium					
Lake	20	413.91	27.43	372.92	475.83
I-O	20	30.25	26.70	−22.40	87.00
dM	20	−4.76	29.52	−76.80	34.10
Sodium					
Lake	20	1922.50	222.72	1610.30	2227.83
I-O	20	27.07	146.61	−251.70	427.90
dM	20	31.15	96.05	−102.80	213.00
Hydrogen ion					
Lake	10	272.58	48.01	207.42	347.14
I-O	20	9521.75	1483.16	6763.60	12,890.20
dM	20	−5.73	53.02	−123.10	103.70
Chloride					
Lake	20	2607.21	392.74	1956.60	3111.03
I-O	20	−286.71	191.51	−617.00	249.00
dM	20	55.78	163.09	−268.90	361.20
Sulfate					
Lake	20	4009.01	449.41	3413.79	4810.23
I-O	20	487.66	382.41	−108.10	1555.90
dM	20	−75.1	325.84	−755.10	593.6
ANC					
Lake	10	71,672.42	5494.79	62,750.56	78,207.84
I-O	9	11,936.11	4794.24	4886.60	17,273.30
dM	10	917.00	10,156.42	−17,817.00	19,505.00

(continued)

TABLE 3-10 (CONTINUED).

Chemical	Number of years	Average	Standard deviation	Minimum	Maxiumum
Phosphate					
Lake	20	6.04	4.39	0.87	17.71
I–O	20	2.66	1.98	0.10	7.90
dM	20	−0.16	6.90	−21.90	12.60
Nitrate					
Lake	20	42.94	19.69	19.86	89.23
I–O	20	275.59	39.20	221.60	344.00
dM	20	0.65	28.67	−43.20	51.90
Ammonium					
Lake	20	20.28	12.66	8.64	50.21
I–O	20	38.29	12.04	18.40	59.10
dM	20	−1.38	17.84	−51.90	30.20
Silicate					
Lake	20	1911.96	247.35	1436.15	2429.98
I–O	20	2847.45	594.69	1973.70	4311.30
dM	20	45.39	463.52	−658.80	824.50
DOC					
Lake					
I–O	4	47.67	165.68	−117.80	272.70
dM	4	0	0	0	0
DIC					
Lake	11	118,215.07	10,409.64	103,962.00	132,425.33
I–O	7	49,733.23	9778.88	38,783.10	63,732.00
dM	9	3325.00	12,384.72	−11,274.00	28,539.00

NOTE: Values for mass in the lake are in kilograms, except for DIC, which is in moles, and ANC and hydrogen ion, which are in equivalents. Values for difference in mass between input and output and change in mass in the lake are kilograms per year, moles per year, or equivalents per year. Mass values are Lake = Mirror Lake, I-O = difference between inputs and outputs, and dM = independently determined change in mass in the lake.

TABLE 3-11. Annual fluxes during driest or wettest years

Chemical	Units	P	NE in	NW in	West in	SWI	GWI	Total in	SWO	GWO	E	Total out	In − out	ΔS	Residual
						Driest year (1988)									
Water	mm	136.9	6.7	143.5	86.9	237.1	104.1	478.1	97.6	396.7	72.6	567.0	−88.9	−1.2	−87.7
Calcium	kg	15.5	74.3	345.6	244.2	664.1	409.0	1088.6	219.3	952.2		1171.5	−82.9	48.1	−131.0
Magnesium	kg	4.5	15.7	75.1	54.1	144.9	82.2	231.6	48.6	216.6		265.2	−33.6	16.4	−50.0
Potassium	kg	7.4	10.5	52.8	50.0	113.3	128.5	249.2	45.0	206.0		251.0	−1.8	33.7	−35.5
Sodium	kg	17.2	175.5	209.0	182.8	567.3	301.3	885.8	185.4	817.3		1002.7	−116.9	134.9	−251.8
Ammonium	kg	38.3	0.3	2.7	2.5	5.5	2.1	45.9	3.1	10.2		13.2	32.7	6.0	26.7
Hydrogen ion	eq	9820.7	9.6	67.4	45.5	122.5	23.3	9966.5	51.3	231.5		282.8	9683.7	0	9683.7
Sulfate	kg	338.6	51.6	806.7	502.9	1361.2	585.9	2285.8	418.1	1816.5		2234.6	51.2	243.6	−192.4
Nitrate	kg	295.9	0.7	13.4	30.7	44.7	0.8	341.4	17.3	44.6		61.8	279.6	4.2	275.4
Chloride	kg	35.5	402.2	141.5	188.3	732.0	114.5	882.0	243.7	1089.7		1333.4	−451.4	116.7	−568.1
Phosphate	kg	5.0	0.2	1.8	2.2	4.2	0.3	9.6	2.1	5.8		7.9	1.7	−21.9	23.5
DOC	kg	130.0	n/a	n/a	n/a	n/a	91.7	n/a	n/a	n/a		n/a	n/a	n/a	n/a
DIC	mol	4790.0	n/a	n/a	n/a	n/a	74,413.6	n/a	n/a	n/a		n/a	n/a	n/a	n/a
ANC	eq	−9820.7	n/a	n/a	n/a	n/a	33,304.0	n/a	n/a	n/a		n/a	n/a	n/a	n/a
Silicate	kg	11.2	79.0	1019.4	746.6	1845.0	1165.6	3021.9	207.8	728.6		936.4	2085.5	−658.8	2744.3

Wettest year (1996)

Water	mm	236.9	20.0	391.8	241.6	653.4	120.9	1011.2	484.3	318.0	58.8	861.1	150.1	4.9	145.2
Calcium	kg	15.9	141.3	661.0	466.5	1268.8	475.1	1759.7	1017.7	667.8		1685.5	74.2	−135.7	209.9
Magnesium	kg	7.0	29.5	140.8	101.6	271.9	95.5	374.4	222.0	146.2		368.2	6.2	−32.1	38.4
Potassium	kg	11.7	23.6	104.1	99.4	227.1	149.3	388.1	217.3	143.3		360.6	27.5	−23.7	51.2
Sodium	kg	44.7	495.1	528.6	519.3	1543.0	349.9	1937.6	1094.4	718.9		1813.3	124.4	−74.5	198.8
Ammonium	kg	56.4	0.1	2.5	1.7	4.3	2.4	63.1	4.9	3.4		8.3	54.9	4.4	50.5
Hydrogen ion	eq	9173.2	153.4	383.8	313.2	850.4	27.1	10,050.7	252.8	168.1		420.9	9629.8	88.4	9541.3
Sulfate	kg	372.5	127.7	1984.6	1129.5	3241.8	680.5	4294.8	2081.5	1374.1		3455.6	839.2	−134.8	974.0
Nitrate	kg	336.4	0.3	5.7	5.0	11.0	1.0	348.4	21.0	17.4		38.4	309.9	8.8	301.1
Chloride	kg	85.1	1004.6	517.5	641.5	2163.6	133.0	2381.6	1570.1	1029.5		2599.6	−218.0	−268.9	50.9
Phosphate	kg	1.9	0.1	0.7	0.9	1.7	0.4	4.0	1.2	0.8		1.9	2.0	−2.2	4.2
DOC	kg	225.1	n/a	n/a	n/a	n/a	106.5	n/a	772.9	477.2		1250.1	n/a	n/a	n/a
DIC	mol	8292.0	6338.0	31,505.0	32,148.0	69,991.0	86,426.8	164,709.8	59,926.0	41,051.8		100,977.8	63,732.0	−3573	67,304.8
ANC	eq	−9173.2	624.0	11,416.0	11,568.0	23,608.0	38,680.5	53,115.3	27,496.0	17,519.2		45,015.2	8100.1	−17817	25,917.2
Silicate	kg	17.7	203.2	2317.2	1692.6	4213.0	1353.8	5584.5	1215.4	774.3		1989.7	3594.9	409.2	3185.7

TABLE 3-12. *Percent contribution to total influx or outflow during driest or wettest years*

Chemical	P	NE in	NW in	West in	GWI	SWO	GWO	E
			Driest year (1988)					
Water	28.6	1.4	30.0	18.2	21.8	17.2	70.0	12.8
Calcium	1.4	6.8	31.7	22.4	37.6	18.7	81.3	
Magnesium	1.9	6.8	32.4	23.4	35.5	18.3	81.7	
Potassium	3.0	4.2	21.2	20.1	51.6	17.9	82.1	
Sodium	1.9	19.8	23.6	20.6	34.0	18.5	81.5	
Ammonium	83.4	0.8	5.8	5.5	4.5	23.0	77.0	
Hydrogen ion	98.5	0.1	0.7	0.5	0.2	18.1	81.9	
Sulfate	14.8	2.3	35.3	22.0	25.6	18.7	81.3	
Nitrate	86.7	0.2	3.9	9.0	0.2	27.9	72.1	
Chloride	4.0	45.6	16.0	21.3	13.0	18.3	81.7	
Phosphate	52.6	1.8	19.2	23.2	3.3	26.9	73.1	
DOC	n/a	n/a	n/a	n/a	n/a	n/a	n/a	
DIC	n/a	n/a	n/a	n/a	n/a	n/a	n/a	
ANC	n/a	n/a	n/a	n/a	n/a	n/a	n/a	
Silicate	0.4	2.6	33.7	24.7	38.6	22.2	77.8	
			Wettest year (1996)					
Water	23.4	2.0	38.7	23.9	12.0	56.2	36.9	6.8
Calcium	0.9	8.0	37.6	26.5	27.0	60.4	39.6	
Magnesium	1.9	7.9	37.6	27.1	25.5	60.3	39.7	
Potassium	3.0	6.1	26.8	25.6	38.5	60.3	39.7	
Sodium	2.3	25.6	27.3	26.8	18.1	60.4	39.6	
Ammonium	89.4	0.2	4.0	2.6	3.8	58.8	41.2	
Hydrogen ion	91.3	1.5	3.8	3.1	0.3	60.1	39.9	
Sulfate	8.7	3.0	46.2	26.3	15.8	60.2	39.8	
Nitrate	96.6	0.1	1.6	1.4	0.3	54.6	45.4	
Chloride	3.6	42.2	21.7	26.9	5.6	60.4	39.6	
Phosphate	47.6	3.0	18.3	22.0	9.1	61.2	38.8	
DOC	n/a	n/a	n/a	n/a	n/a	61.8	38.2	
DIC	5.0	3.8	19.1	19.5	52.5	59.3	40.7	
ANC	−17.3	1.2	21.5	21.8	72.8	61.1	38.9	
Silicate	0.3	3.6	41.5	30.3	24.2	61.1	38.9	

REFERENCES

Asbury, C.E. 1990. The role of groundwater seepage in sediment chemistry and nutrient budgets in Mirror Lake, New Hampshire. Ph.D. dissertation, Cornell University. 275 pp.

Bade, D.L., K. Bouchard, and G.E. Likens. 2009. Algal co-limitation by nitrogen and phosphorus persists after 30 years in Mirror Lake (New Hampshire, USA). *Verh. Internat. Verein. Limnol.* 30 (in press).

Bailey, A.S., J.W. Hornbeck, J.L. Campbell, and C. Eagar. 2003. Hydrometeorological database for Hubbard Brook Experimental Forest: 1955–2000. General Technical Report NE-305, USDA Forest Service, Newton Square, Pennsylvania, 36 pp.

Bailey, S.W., D.C. Buso, and G.E. Likens. 2003. Implications of sodium mass balance for interpreting the calcium cycle of a forested ecosystem. *Ecology* 84(2):471–484.

Barton, C.C., R.H. Camerlo, and S.W. Bailey. 1997. Bedrock Geologic Map of Hubbard Brook Experimental Forest and maps of fractures and geology in roadcuts along Interstate-93, Grafton County, New Hampshire. Sheet 1, Scale 1:12,000; Sheet 2, Scale 1:200. U.S. Geological Survey, Miscellaneous Investigations Series, Map I-2562.

Buso, D.C., G.E. Likens, and J.S. Eaton. 2000. Chemistry of precipitation, tributary water and lake water from the Hubbard Brook Ecosystem Study: A record of sampling protocols and analytical procedures. General Tech. Report NE-275. Newtown Square, Pennsylvania. USDA Forest Service, Northeastern Research Station. 52 pp.

Cogbill, C.V., and G.E. Likens. 1974. Acid precipitation in the northeastern United States. *Water Resour. Res.* 10(6):1133–1137.

Cole, J.J., N.F. Caraco, G.W. Kling, and T.W. Kratz. 1994. Carbon dioxide supersaturation in the surface waters of lakes. *Science* 265:1568–1570.

Dillon, P.J., and F.H. Rigler. 1974. A test of a simple nutrient budget model predicting the phosphorus concentration in lake water. *Journal of the Canadian Fisheries Research Board of Canada* 31:1771–1778.

Driscoll, C.T., G.B. Lawrence, A.J. Bulger, T.J. Butler, C.S. Cronan, C. Eagar, K. Fallon Lambert, G.E. Likens, J.L. Stoddard, and K.C. Weathers. 2001. Acidic deposition in the northeastern United States: Sources and inputs, ecosystem effects, and management strategies. *BioScience* 51(3):180–198.

Gbondo-Tugbawa, S.S., and C.T. Driscoll. 2002. Evaluation of the effects of future controls on sulfur dioxide and nitrogen oxide emissions on the acid-base status of a northern forest ecosystem. *Atmos. Environ.* 36:1631–1643.

Gerhart, D.Z., and G.E. Likens. 1975. Enrichment experiments for determining nutrient limitation: Four methods compared. *Limnol. Oceanogr.* 20(4):649–653.

Giblin, A.E., G.E. Likens, D. White, and R.W. Howarth. 1990. Sulfur storage and alkalinity generation in New England lake sediments. *Limnol. Oceanogr.* 35(4):852–869.

Harte, P.T., and T.C. Winter. 1995. Simulations of flow in crystalline rock and recharge from overlying glacial deposits in a hypothetical New England setting. *Ground Water* 33(6):953–964.

Helsel, D.R., and R.M. Hirsch. 1997. *Statistical Methods in Water Resources.* Studies in Environmental Science 49. New York: Elsevier. 529 pp.

Kaushal, S.S., P.M. Groffman, G.E. Likens, K.T. Belt, W.P. Stack, V.R. Kelly, L.E. Band, and G.T. Fisher. 2005. Increased salinization of fresh water in the northeastern United States. *Proc. National Academy of Sciences* 102(38):13517–13520.

Krabbenhoft, D.P., and K.E. Webster. 1995. Transient hydrogeological controls on the chemistry of a seepage lake. *Water Resour. Res.* 31(9):2295–2305.

LaBaugh, J.W. 1991. Spatial and temporal variation in the chemical characteristics of ground water adjacent to selected lakes and wetlands in the North Central United States. *Verhandlungen Internationale Vereinigung Limnologie* 24:1588–1594.

Likens, G.E. (ed). 1985. *An Ecosystem Approach to Aquatic Ecology: Mirror Lake and Its Environment.* New York: Springer-Verlag. 516 pp.

Likens, G.E., and F.H. Bormann. 1974. Acid rain: A serious regional environmental problem. *Science* 184(4142):1176–1179.

Likens, G.E., and F.H. Bormann. 1995. *Biogeochemistry of a Forested Ecosystem, Second Edition.* New York: Springer-Verlag. 159 pp.

Likens, G.E., F.H. Bormann, and N.M. Johnson. 1972. Acid rain. *Environment* 14(2):33–40.

Likens, G.E., and D.C. Buso. 2006. Variation in streamwater chemistry throughout the Hubbard Brook Valley. *Biogeochemistry* 78:1–30.

Likens, G.E., D.C. Buso, and T.J. Butler. 2005. Long-term relationships between SO_2 and NO_x emissions and SO_4^{2-} and NO_3^- concentration in bulk deposition at the Hubbard Brook Experimental Forest, New Hampshire. *J. Environ. Monitoring* 7(10):964–968.

Likens, G.E., D.C. Buso, and J.W. Hornbeck. 2002. Variation in chemistry of stream water and bulk deposition across the Hubbard Brook Valley, New Hampshire, USA. *Verh. Internat. Verein. Limnol.* 28(1):402–409.

Likens, G.E., C.T. Driscoll, and D.C. Buso. 1996. Long-term effects of acid rain: Response and recovery of a forest ecosystem. *Science* 272:244–246.

Likens, G.E., C.T. Driscoll, D.C. Buso, M.J. Mitchell, G.M. Lovett, S.W. Bailey, T. G. Siccama, W.A. Reiners, and C. Alewell. 2002. The biogeochemistry of sulfur at Hubbard Brook. *Biogeochemistry* 60(3):235–316.

Likens, G.E., J.S. Eaton, and N.M. Johnson. 1985. Mirror Lake—physical and chemical characteristics: B. Physical and chemical environment. In G.E. Likens (ed.). *An Ecosystem Approach to Aquatic Ecology: Mirror Lake and Its Environment* (pp. 89–108). New York: Springer-Verlag.

Likens, G.E., and R.E. Moeller. 1985. Chemistry. In G.E. Likens (ed.). *An Ecosystem Approach to Aquatic Ecology: Mirror Lake and Its Environment* (pp. 392–410). New York: Springer-Verlag.

Rosenberry, D.O., P.A. Bukaveckas, D.C. Buso, G.E. Likens, A.M. Shapiro, and T.C. Winter. 1999. Movement of road salt to a small New Hampshire lake. *Water, Air, and Soil Pollution* 109:179–206.

Stelzer, R.S., and G.E. Likens. 2006. Effects of sampling frequency on estimates of dissolved silica export by streams: The role of hydrological variability and concentration-discharge relationships. *Water Resour. Res.* 42(7):W07415

Wentz, D.A., W.J. Rose, and K.E. Webster. 1995. Long-term hydrologic and biogeochemical responses of a soft water seepage lake in north central Wisconsin. *Water Resour. Res.* 31(3):199–212.

Wetzel, R., and G.E. Likens. 1991. *Limnological Analyses.* New York: Springer-Verlag. 391 pp.

Winter, T.C. 1981. Numerical simulation analysis of the interaction of lakes and ground water. U.S. Geological Survey Professional Paper 1001, 45 pp.

Winter, T.C. 1985. Approaches to the study of lake hydrology. In G.E. Likens (ed.). *An Ecosystem Approach to Aquatic Ecology: Mirror Lake and Its Environment* (pp. 128–135). New York: Springer-Verlag.

4

EVALUATION OF METHODS AND UNCERTAINTIES IN THE WATER BUDGET

THOMAS C. WINTER AND
DONALD O. ROSENBERRY

Measurements of the volumes of water moving to, stored in, and moving from a lake are subject to some degree of uncertainty. The uncertainties in measuring water flows result from less than optimal placement of field instruments, the limited accuracy and reliability of field instruments, missing data, and the challenge of areally interpreting point data from field measurements. The methods used to quantify the water budget components of Mirror Lake are described briefly in chapter 2. This chapter presents more detail on how those values were derived, including the assumptions that went into the calculations and the uncertainties inherent in the values.

WATER STORAGE IN THE LAKE

Mirror Lake is a relatively small lake that has a simple geometry; the lakebed slopes sharply downward from the near-shore littoral zone to a gently sloping abyssal plain (Fig. 1-2). As a result of the many depth measurements made to define the shape of the lakebed in this detail, the volume of the lake can be determined relatively accurately if the lake stage is measured accurately. During this study, the lake stage was measured continuously using an analog strip-chart recorder and float system, where the charts could easily be read to within 2 mm. However, the trace of the lake stage on the chart can be highly variable under windy conditions making it difficult to determine daily average stage for some days. Therefore, we estimate the error in changes in lake storage to be about 10 percent.

PRECIPITATION

Uncertainties in precipitation values can result from improper placement of the gages, limits in the precision of the gages, and limits in the density of gages used to extrapolate precipitation over the study area. With respect to placement, precipitation gages located in forested areas are subject to encroaching vegetation, especially during long-term studies. A cone of open air should surround a precipitation gage so that the height of nearby objects is preferably less than, but no more than equal to, the distance of that object from the rain gage (i.e., the angle of an opening of the collector to the sky should be at least 45 degrees from horizontal). As trees continue to grow over time, it is important to trim vegetation to maintain the proper opening to the sky to minimize any temporal bias associated with reduced precipitation catch by the gage. For the 20 years of this study, the openings in the forest where the gages were placed were sufficiently large to avoid this issue.

To overcome the problem of catch efficiency of the gages related to wind, the precipitation gages used in this study were equipped with alter shields. These shields encircle the gage and are designed to disrupt the wind flow passing over the gage, causing the rain and snow to fall more vertically into the gage (Fig. 2-9) and thereby increasing their accuracy. Based on a review of the literature related to rain gage errors (Winter 1981), it is estimated that uncertainty in monthly measurements of precipitation related to the gages themselves probably is about 2 to 4 percent.

Averaging the precipitation measured at two gages some distance from the lake is a potential source of uncertainty for the water budget of Mirror Lake. Precipitation was not measured at the lake; one gage is located about 0.4 km west of the lake, and the other is about 0.5 km southeast of the lake. For the 20-year period, the average difference between the two gages was 1.3 mm, and the standard deviation was 2.0 mm. Because of this small difference, it is believed that the average values of the precipitation measured by these two gages adequately represents the precipitation falling on the lake. Considering error associated with the gages themselves and with the method used to areally distribute those point data, the overall error for monthly precipitation input to Mirror Lake is estimated to be about 5 percent.

EVAPORATION

One of the goals of the water budget study of Mirror Lake was to measure each component of the hydrologic system interacting with the lake as accurately as possible. During the late 1970s, when hydrologic studies of Mirror Lake began, the Bowen ratio energy budget method (Harbeck et al. 1958) was considered to be the best for accurate long-term monitoring of evaporation. Although the eddy correlation method is generally accepted as being more accurate, only in recent years have the instruments needed to use this method become robust enough to be used for long-term monitoring. Even though the energy budget was the best method available, it requires a great deal of instrumentation (described in chapter 2), frequent field checks of those instruments, careful quality assurance of the data, and regular thermal surveys of the lake water.

During the course of the study, new sensors became available that provided either more accurate or more convenient measurement of variables. Several sensors were used for measuring vapor pressure, for example, and, as described in chapter 2, considerable effort was made to minimize bias during each transition to a new sensor. Fortunately, the parameters that have the greatest effect on energy budget determinations—shortwave and longwave radiation—were measured during the entire period with the same high-quality sensors.

Because of concerns regarding the costs of maintaining long-term accuracy of the field data, the original plan for long-term monitoring of evaporation at Mirror Lake was to use a simpler method that would require fewer and very robust instruments. The method originally intended to be used was the mass transfer method, which required measurement only of air and water surface temperatures, vapor pressure, and wind speed. A major drawback of the mass transfer method, however, is that all these parameters need to be measured at the center of the lake; hence, a raft was needed (Fig. 2-10). Also, a mass transfer coefficient needs to be determined that is unique to each lake, and it needs to be determined by calibration against an independent measurement of evaporation. Because of the need both for accurate estimates of evaporation and for cost-effectiveness over the long term, the decision was made to calibrate the mass transfer coefficient against the energy budget method for a period of six years (1982–1987), and then to monitor evaporation over the long term using the mass transfer method. However, a comparison of 14 evaporation methods with evaporation determined by the energy budget method (Rosenberry et al. 2007) indicated that a number of methods were better than the mass transfer method (Table 4-1). As a result of this comparison, the Priestley-Taylor method (Stewart and Rouse 1976) was selected to determine evaporation from Mirror Lake for the 20-year study. (These are the values presented in chapter 2.)

Even though the Priestley-Taylor method requires more instrumentation than the mass transfer method, including use of relatively expensive radiometers and labor-intensive thermal surveys, it does not require data for wind speed or vapor pressure. Not needing these data made it unnecessary to have a raft in the middle of the lake. A two-year comparison of measuring air and water surface temperatures at the raft and at a boom extending over the water several meters from shore indicated little difference in these variables between the two locations, further negating the necessity of a raft. The only major drawback of the Priestley-Taylor method is that it requires periodic thermal surveys of the lake water. However, this expense can be mitigated with little additional error for small lakes having a simple conical basin shape like Mirror Lake because a single, centrally located thermal survey was found to adequately determine the heat stored in the lake (Rosenberry et al. 1993).

The estimated uncertainty in monthly evaporation determined for Mirror Lake is about 15 percent. This estimate is based on studies by Harbeck et al. (1958) and Gunaji (1968), which indicated that monthly evaporation determined by the energy budget method had uncertainty of about 10 percent. The additional uncertainty involved in estimating monthly evaporation using the Priestley-Taylor method is estimated to be about 5 percent.

SURFACE WATER

TRIBUTARY INFLOW

Parshall flumes were used to measure discharge into Mirror Lake from the three tributaries throughout the 20-year study (Fig. 2-12). These types of flumes have been calibrated in laboratory studies such that the discharge through them can be calculated knowing only the water level (hydraulic head) in the inlet section. To measure the head, a stilling well is placed outside the flume adjacent to the inlet section and is connected to that section by a short pipe, allowing free movement of water between the flume and the stilling well. A float and recorder system is then placed over the stilling well to continuously record head changes in the flume. This simple setup is very reliable; the main problems usually are related to (1) a malfunction of the clock that drives the recorder pen across the chart, (2) sediment obstructing the connector pipe between the flume and stilling well, (3) occasional overtopping of the flume during periods of very high runoff, and (4) operation of the flumes during winter when temperatures are cold enough for ice to form in the stilling well. This last problem was dealt with by building shelters around the flumes and providing a heat source to prevent ice formation. Fortunately, because of the similar setting of Mirror Lake's tributaries, discharge among them is highly correlated. Therefore, periods of missing streamflow data at one tributary were filled in by regression equations developed by correlation with another.

According to the manufacturer of the Parshall flumes, the error in measured discharge is about 2 percent. However, field conditions are seldom as ideal as laboratory conditions. Although Parshall flumes are designed to be self-cleaning because of their flow hydraulics, sediment sometimes is deposited in their inlet section; alternatively, a large object, usually wood, sometimes gets stuck upstream from the throat section. Therefore, it is necessary to inspect the flumes regularly to make sure they are operating properly. In addition, peak-flow data are lost when the flumes are overtopped. Because of these occasional problems with the flume and recorder system, a more realistic estimate of the error in discharge for the Mirror Lake tributary streams is about 5 percent.

SURFACE OUTFLOW

Surface outflow from Mirror Lake occurs when the lake stage rises above the top boards of the dam (Fig. 2-8). In addition, a minor amount of water is lost by seepage between the boards that compose the dam. Before 1990, surface outflow was determined by assuming that the flow over the top board of the dam could be approximated by using a formula for a broad-crested weir. A wooden board is a less-than-ideal structure for accurately measuring stream discharge; therefore, to improve the accuracy of measuring the surface outflow from the lake, a Parshall flume was constructed about 10 m downstream from the outlet dam in late 1989. The flume needed to be sufficiently large (it has a throat section 61 cm wide) to accommodate the large discharges that sometimes flow from the lake. However, having such a large throat results in less accuracy in measurement of low flows. To measure low flows accurately, a removable weir plate is attached to the outlet end of the flume during times of low flow (Fig. 2-13). With the weir plate installed, the flume itself encloses part of the pool that is needed for a weir to work properly, and the same stilling well and recorder system can be used. The field person simply needs to note when the weir plate is attached or removed so the correct formula (flume or weir) can be used.

The flume came online in spring of 1990; therefore, it was necessary to develop a method to back-calculate surface outflow for the period from January 1981 to May 1990. To accomplish this, a statistical relation was first developed between lake stage in inches above the top of the boards and discharge measured at the flume. The relation is a three-part equation.

For lake stages higher than 2.15 inches (5.46 cm) above the boards, the equation is linear:

$$Q = 14.17x - 30.25 \ (R^2 = 0.93) \qquad (4\text{-}1)$$

where

Q is in cubic feet per second, and

x is lake stage, in inches.

For lake stages between 2.0 (5.08 cm)(top of the boards) and 2.15 inches, the equation is nonlinear:

$$Q = 78.78 - 77.91x + 19.27x^2 \ (R^2 = 0.56) \qquad (4\text{-}2)$$

For lake stages below the top of the boards, a constant outflow of 0.030 ft^3 s^{-1} was used to quantify the seepage between the boards. The relatively low R^2 for lake stages between 2.0 and 2.15 inches is an indication of the complex hydraulics of flow near the top of the boards. Unfortunately, the lake stage is between 2.0 and 2.15 inches above the boards more often than it is higher than 2.15 inches. The lesser accuracy during these periods is partly offset, however, by the fact that far more water leaves the lake at the higher stages, when the measurements have greater accuracy.

Comparison of discharge values calculated by the three-part equation and the discharge measured at the flume for June through December 1990 indicated a mean average daily difference of 0.057 ft^3 s^{-1} and a standard deviation of 0.240 ft^3 s^{-1}. These data indicate that the surface outflow discharge calculated for the 1980s using the three-part equation has greater error than the values calculated from the flume data following May 1990. The error in measurements of daily average outflow discharge for January 1981 through May 1990 is estimated to be 10 to 15 percent. The error associated with measurements of daily average outflow discharge following May 1990 using the flume (or weir) is estimated to be about 5 percent.

GROUND WATER

The exchange of Mirror Lake water with ground water generally is complex because of the variability in the texture and composition of the underlying geologic deposits. As a result, it is difficult to determine inflow seepage from and outflow seepage to ground water; it generally is prudent to use several methods to arrive at acceptable values. For this study, three methods were used: (1) the analytical Darcy method, (2) stable isotope ratios of oxygen, and (3) a groundwater simulation model.

FLUXES OF GROUND WATER INTERACTING WITH THE LAKE

The Analytical Darcy Method

To use the analytical Darcy method, the following equation is solved:

$$Q = KIA \qquad (4\text{-}3)$$

where

K is the hydraulic conductivity of the geologic materials, in m day^{-1},

I is the hydraulic gradient between a well and the lake, unitless, and

A is a vertical plane at the shoreline and is the cross-sectional area through which water must pass to flow from ground water to the lake or from lake water to ground water, in m^2.

Hydraulic conductivity has the greatest uncertainty of the parameters that compose the groundwater flow equation. The main reason for this is the large degree of heterogeneity associated with most glacial deposits. Glacial till is a heterogeneous mixture of clay through boulder-sized rocks that has no structure and generally has low hydraulic conductivity (Fig. 2-1). The sand and gravel deposit on the south side of the lake was deposited by flowing water, which results in a complex stratified structure having a wide range of particle sizes over short distances, but generally has high hydraulic conductivity. As described in chapter 2, hydraulic conductivity was determined by single-well hydraulic "slug" tests. This method tests the hydraulic conductivity of only a small volume of geologic deposits in the vicinity of the well. Therefore, to obtain an estimate of the average hydraulic conductivity of the geologic deposits as a whole, many wells need to be tested. Although tests at dozens of wells indicated a range in hydraulic conductivity of more than an order of magnitude for the till deposits, indicating the great heterogeneity of the deposits, the average was about 0.3 ft day^{-1}. The range in hydraulic conductivity for the sand and gravel deposits was smaller, and the average was about 15 ft day^{-1}.

Values for *I* were determined by calculating the difference in water levels between a well and the lake, divided by the distance between the well and the lake. This procedure required all wells to be surveyed to a common datum for proper reference to lake stage. The water level in a well and the water level of the lake can be measured very accurately, to within a few millimeters. However, gradients measured between an index well and the lake are assumed to be uniform for any given segment of shoreline shown on Fig. 4-1. There undoubtedly is error associated with this assumption, but it cannot be quantified without drilling many more wells to determine the variability in gradients. The reasonableness of the assumption, however, can be determined by comparing groundwater fluxes calculated by the analytical Darcy method and fluxes determined by a numerical groundwater flow model (discussed later in this chapter).

The *A* term was determined by measuring the thickness of the glacial deposits along the lakeshore using geophysical methods and test drilling. The lakeshore was then divided into segments based on the position of "index wells," which are those used for determining hydraulic gradient (Fig. 4-1). The length of segment multiplied by the thickness of glacial deposits determined *A* for each segment. Given

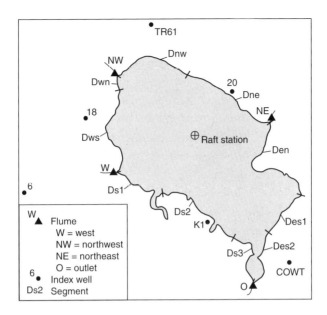

FIGURE 4-1. Segments of shoreline across which groundwater flux was calculated using the analytical Darcy method. Also shown are the index wells used for calculating hydraulic gradients for the various segments.

the characteristics of groundwater flow near a surface water body, the flow lines are most likely to cross lines of equal hydraulic head at right angles through a vertical plane along the shoreline. Therefore, the thickness of the groundwater system interacting with the lake needs to be determined, along with the length of each shoreline segment associated with an index well. At Mirror Lake, seismic geophysical surveys were used to determine the thickness of the glacial deposits around the perimeter of the lake. These measurements ranged from 0 to more than 30 m, and are estimated to have error of about 10 percent (John Lane, U.S. Geological Survey, oral communication 2007). Of the three terms on the right side of Equation 4-3, A probably has the least uncertainty.

Groundwater fluxes were calculated on a monthly basis. Because K and A are constant, monthly fluxes were determined using the average I for each month.

The water budgets calculated using the analytical Darcy method for Mirror Lake used the hydraulic conductivities mentioned above for determining gains from and losses to ground water. However, evidence from carbon budgets (Jon Cole and Nina Caraco, Institute of Ecosystem Studies, oral communication 2005) indicated that inflow from ground water needed to be greater in order for the carbon budget to balance. Furthermore, Schulze-Makuch et al. (1999) indicated that hydraulic conductivity determined by single-well slug tests may underestimate the bulk hydraulic conductivity of an entire geologic formation. Based on this information, it was decided to examine two other methods for determining the groundwater inflow to Mirror Lake: (1) the mass balance of the stable isotopes of oxygen, and (2) a numerical groundwater flow model (which had already been done by Tiedeman et al. 1997).

Stable Isotope Ratios for Oxygen

The stable isotope ratio for oxygen in water $\delta\ ^{18}O$ can be used to estimate groundwater inflow to a lake when the stable oxygen-isotope ratios are known for lake water, tributaries, ground water entering the lake, and precipitation. The method also requires knowing the oxygen isotope composition of the water vapor leaving the lake as evaporation. Samples of lake water, tributary water, and ground water from a number of wells throughout the watershed were collected in 2004, and samples of precipitation were collected over a period of six months that same year. The samples were collected in 250 ml glass bottles, and the isotopes of oxygen were analyzed at the Isotope Laboratory at the University of Waterloo in Canada.

The water budget equation for a lake at hydrologic steady state is

$$dV/dt = GWI + P + SWI$$
$$- GWO - E - SWO = 0 \qquad (4\text{-}4)$$

where

dV is the change in lake volume,

dt is change in time,

GWI is the volume of groundwater seepage to the lake,

P is the volume of atmospheric precipitation on the lake,

SWI is the volume of surface water discharge to the lake,

GWO is the volume of lake water seepage to ground water,
E is the volume of evaporation, and
SWO is the volume of surface water discharge from the lake.
(All values are in m^3.)

The isotope mass balance for a lake at steady state is

$$d(V\delta_L)/dt = GWI\ \delta_{GWI} + P\ \delta_P + SWI\ \delta_{SWI}$$
$$- GWO\ \delta_L - E\ \delta_E - SWO\ \delta_L = 0 \qquad (4\text{-}5)$$

where

L is the lake water, and
the δ values are the $\delta\ ^{18}O$ isotopic ratio compositions of the water (per mil) collected from the indicated components of the hydrologic system.

Evaporated water vapor was not collected for isotopic analysis; δ_E was calculated as described later in this section. Water leaving the lake either as seepage to ground water or as surface outflow was assumed to have the same oxygen isotope composition as the lake.

At hydrologic steady state, Equation 4-5 can be rearranged to solve for the volume of groundwater discharge to the lake:

$$GWI = [P\ (\delta_L - \delta_P) + SWI\ (\delta_L - \delta_{SWI})$$
$$+ E\ (\delta_E - \delta_L)] / (\delta_{GWI} - \delta_L) \qquad (4\text{-}6)$$

Also, at hydrologic steady state, Equation 4-5 can be rearranged to solve for the ratio of total outflow to evaporation:

$$O/E = [(SWI_{nw}/E)(\delta_{GWI} - \delta_{SWInw}) + (SWI_w/E)(\delta_{GWI} - \delta_{SWI})$$
$$+ (SWI_{ne}/E)(\delta_{GWI} - \delta_{SWIne}) + (P/E)(\delta_{GWI} - \delta_P)$$
$$+ \delta_E - \delta_{GWI}] / (\delta_{GWI} - \delta_L) \qquad (4\text{-}7)$$

where

O is total outflow (ie, $SWO + GWO$),
nw is the Northwest Tributary,
w is the West Tributary, and
ne is the Northeast Tributary.

Therefore,

groundwater outflow = total outflow − surfacewater outflow.

The calculation of δ_E comes from Craig and Gordon (1965) and is shown by Krabbenhoft et al. (1990) as

$$\delta_E = (\alpha^*\delta_L - h\,\delta_A - e) / (1-h + 10^{-3}\,\Delta e) \qquad (4\text{-}8)$$

where

α^* is the equilibrium isotopic fractionation factor at the temperature of the air-water interface.

This term was calculated from Majoube (1971) as

$$\alpha^* = exp((1137^*(T^{**}-2)) - (0.4156^*(T^{**}-1)) - 0.00207) \qquad (4\text{-}9)$$

where

T is the lake surface temperature, in degrees Kelvin,

δ_L is the isotopic value of the lake water,

h is relative humidity,

δ_A is the isotopic value of local atmospheric water vapor,

e is the total fractionation factor [$e = 1000(1-\alpha^*) + \Delta e$], and

Δe is the diffusion controlled or kinetic fractionation factor, which is a function of relative humidity and is calculated as $\Delta e = K(1-h)$, where $K(^{18}O) = 14.3$ from wind tunnel experiments done by Gilath and Gonfiantini (1983).

It is best to measure δ_A directly, but it can be approximated from the value for precipitation. Krabbenhoft et al. (1990) collected water vapor for isotopic analysis and found that using the value for precipitation as a substitute is valid.

Results of the oxygen-isotope method, where Equation 4-6 was used to calculate groundwater discharge to the lake, are shown in Table 4-2. The table shows the volumes of water fluxes to and from the lake as well as the isotopic composition of those waters. Groundwater discharge to the lake estimated by the isotope mass balance method is approximately twice as large as the volume calculated from the analytical Darcy method (Equation 4-3).

NUMERICAL MODEL OF INFLOW FROM THE GROUNDWATER BASIN

Tiedeman et al. (1997) developed a three-dimensional steady-state numerical model of the Mirror Lake groundwater basin as a means to better understand the movement of ground water throughout the basin and its interaction with surface water, including Mirror Lake. The model domain was bounded on the west by Paradise Brook, on the north by Leeman's Brook, on the east by the Pemigewasset River, and on the southern and northwestern corners by topographic highs. All of these represent no-flow boundaries. The model was discretized into 89 rows, 85 columns, and 5 layers. Only the part of the model covering Mirror Lake and its watersheds is shown in Fig. 4-2. (See Tiedeman et al. for a map of the entire model domain.) In a numerical simulation, the groundwater flow equation

$$\frac{\partial}{\partial x}\left(K_H \frac{\partial h}{\partial x}\right) + \frac{\partial}{\partial y}\left(K_H \frac{\partial h}{\partial y}\right) + \frac{\partial}{\partial z}\left(K_V \frac{\partial h}{\partial z}\right) = 0 \qquad (4.10)$$

where

h is the hydraulic head (L),

K_H is the horizontal hydraulic conductivity (L T^{-1}),

K_V is the vertical hydraulic conductivity (L T^{-1}),

x, y are the Cartesian coordinates in the horizontal directions (L), and

z is the Cartesian coordinate in the vertical direction (L)

is solved for each cell of the entire domain repeatedly until mass balance is achieved and there is a very small difference in head distribution and tributary base flow between successive solutions. The solution must also meet calibration targets, which for the Mirror Lake model were the base flow of the three tributaries, as well as the measured groundwater heads in numerous piezometers scattered throughout the area near Mirror Lake (Fig. 4-2). As part of the model solution, groundwater discharges to and from surface water bodies are also calculated. As a result of the finer discretization of the model grid (Fig. 4-2), there are many more segments of the Mirror Lake shoreline for which flux to the lake is calculated than the number of segments used for the original analytical Darcy calculations (Fig. 4-1).

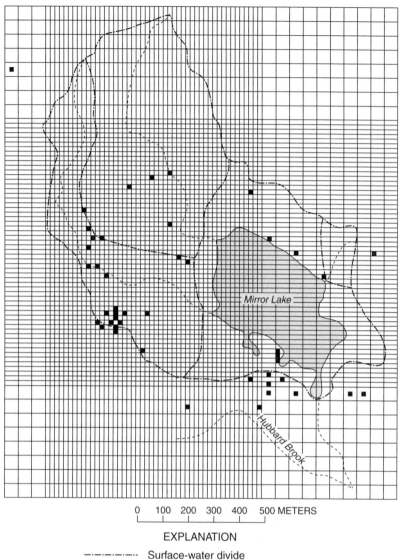

FIGURE 4-2. Discretization grid for a numerical simulation model of the groundwater basin of Mirror Lake. Streams and observation wells used as calibration targets also are shown. (From Tiedeman et al. 1997.)

The simulations made by Tiedeman et al. (1997) also made use of a computer code that estimates parameters that were not measured, such as groundwater recharge and hydraulic conductivity. The model-estimated hydraulic conductivity for glacial till ranged from 0.48 to 0.77 ft day^{-1}, which is 1.7 to 2.7 times greater (average of 2.2 times greater) than the hydraulic conductivity used for the original analytical Darcy calculations (0.3 ft day^{-1}).

The water budget components derived from the numerical analysis of the Mirror Lake groundwater basin indicates that the long-term average groundwater discharge from the glacial deposits to Mirror Lake was 133,000 m^3 yr^{-1}. This volume is 2.8 times greater than the long-term average of 47,000 m^3 yr^{-1} calculated from the analytical Darcy method.

The alternative methods used to determine groundwater inflow to Mirror Lake indicated that the values calculated from the analytical Darcy method needed to be increased. The oxygen isotope method indicated that these values be multiplied by a factor of 2. The parameter-estimated hydraulic conductivity determined by the numerical groundwater model indicated that the values calculated from the analytical Darcy method be multiplied by a factor of 2.2, if it is assumed that the difference in groundwater inflow is related only to hydraulic conductivity. Groundwater flux to the lake determined by the numerical groundwater model indicated that the values calculated from the analytical Darcy method be multiplied by a factor of 2.8. The difference between the adjustments suggested by the hydraulic conductivity alone and that suggested by the numerically simulated flux probably is related to differences in hydraulic gradient. As mentioned earlier, the model had been discretized such that groundwater flux was calculated for many more (thus smaller and more accurate) segments of shoreline than were used for the analytical Darcy calculations, which assumed average gradients for large segments of shoreline (compare Figs. 4-1 and 4-2). Considering all methods, and the uncertainties in each, it was decided to increase the original values of groundwater inflow to Mirror Lake by a factor of 2.4. Furthermore, because all other terms in Mirror Lake's water budget are fairly well constrained, except for lake seepage to ground water, it was decided that the additional volume represented by adjusting groundwater inflow by 2.4 times (47,000 × 2.4 = 113,000 − 47,000 = 66,000 m^3 yr^{-1})

needed to be added to the lake's loss to ground water (*GWO*). Thus, the lake seepage to ground water also was increased by 66,000 m^3 yr^{-1}, which resulted in a multiplier of 1.234 for the lake seepage outflow values. The values of groundwater inflow (*GWI*) and lake seepage outflow (*GWO*) discussed in other chapters of this book reflect these adjustments.

Using hydraulic conductivity values greater than those determined by single-well slug tests also is consistent with the findings of Schulze-Makuch et al. (1999). Their study examined the scale dependency of determining hydraulic conductivity and found that the larger the volume of rock tested, the larger are the values of hydraulic conductivity determined. For example, for glacial outwash in Wisconsin, hydraulic conductivities determined by single-well slug tests, where the deposits in the immediate vicinity of the well are tested, were about an order of magnitude greater than hydraulic conductivities determined for cores in laboratory permeameters. Furthermore, hydraulic conductivities determined by pumping tests at the field scale and by numerical modeling were several orders of magnitude greater than hydraulic conductivities determined by single-well slug tests.

The numerical groundwater model was of steady-state conditions, which indicates that it is representative only of long-term average conditions. For example, the targets for calibration of the model were the long-term averages of the various tributary base flows and groundwater hydraulic heads used. Therefore, the values of the parameters used to determine the adjustment needed for groundwater inflow and lake seepage outflow also were the 20-year average values. However, this study is concerned with monthly and annual values of Mirror Lake's water budget. The calculations of groundwater exchange with the lake using the analytical Darcy method are based on actual measurements of groundwater and lake water heads; therefore, they represent the actual variability of the month-to-month hydraulic gradients. To capture this variability, all monthly values were adjusted using the multipliers for groundwater inflow and lake seepage outflow mentioned above. In a sense, this approach is similar to saying that the adjustment is based on a constant, such as hydraulic conductivity. We believe the adjustment is justified on this basis because the multiplier of 2.4 falls within the range of hydraulic conductivity estimated by the numerical groundwater model. Considering all of the factors involved in trying to determine the exchange of ground water with Mirror Lake, the error in the groundwater values is estimated to be about 25 percent.

TABLE 4-1. *Regression R^2, slope, and offset coefficients for model output versus Bowen ratio energy budget (BREB) values, and percent of monthly periods that alternate evaporation values are within 5, 10, and 20 percent of BREB values*

Alternate method	R^2 regressed against BREB	Regression slope coeff. vs. BREB	Regression offset vs. BREB	Results within 5% of BREB	Results within 10% of BREB	Results within 20% of BREB
Priestley-Taylor	0.97	1.24	−0.35	24%	38%	97%
deBruin-Keijman	0.97	1.24	−0.27	14%	30%	95%
Penman	0.96	1.15	−0.13	30%	62%	92%
Brutsaert-Stricker	0.92	1.33	−0.57	22%	38%	65%
Papadakis	0.57	0.79	0.87	11%	43%	60%
Thornthwaite	0.73	1.04	−0.07	38%	46%	59%
Mass transfer	0.55	0.77	0.53	14%	35%	57%
Stephens-Stewart	0.74	1.13	−0.69	14%	22%	57%
Makkink	0.74	1.18	−0.42	14%	27%	54%
deBruin	0.38	0.60	0.50	5%	19%	49%
Hamon	0.71	1.31	−0.57	14%	19%	46%
Ryan-Harleman	0.60	1.02	0.64	3%	16%	22%
Jensen-Haise	0.74	1.83	−1.21	3%	11%	19%
Blaney-Criddle	0.73	1.71	−0.66	3%	3%	19%

SOURCE: From Rosenberry et al. 2007.
NOTE: $n = 37$ except for Papadakis model, where $n = 35$.

TABLE 4-2. *Water budget of Mirror Lake calculated from the oxygen-isotope method*

Budget component	Volume, in m³ per year	$\delta^{18}O$
Ground water in (GWI)	94,854	−9.12
Precipitation (P)	181,981	−8.71
NW tributary (SWI_{NW})	248,049	−8.76
W tributary (SWI_W)	157,296	−8.71
NE tributary (SWI_{NE})	12,140	−8.69
Total in	694,320	
Evaporation (E)	76,646	−27.19
Ground water out (GWO)	366,991	−6.50
Lake surface outlet (SWO)	250,683	−6.50
Total out	694,320	

REFERENCES

Craig, H., and L.I. Gordon. 1965. Deuterium and oxygen 18 variations in the ocean and marine atmosphere. In *Stable Isotopes in Oceanographic Studies and Paleotemperatures* (pp. 9–130). Pisa: Spoleto, Consiglio Nazionale delle Ricerche.

Gilath, C., and R. Gonfiantini. 1983. Lake dynamics. In *Guidebook on Nuclear Techniques in Hydrology* (pp. 129–161). Vienna: International Atomic Energy Agency Technical Report Series 91.

Gunaji, N.N. 1968. Evaporation investigations at Elephant Butte Reservoir in New Mexico. In *Geochemistry, Precipitation, Evaporation, Soil Moisture, Hydrometry* (pp. 308–325). General Assembly of Bern, International Association of Scientific Hydrology Publication 78.

Harbeck, G.E., Jr., M.A. Kohler, and G.E. Koberg. 1958. Water-loss investigations, Lake Mead studies. U.S. Geological Survey Professional Paper 298, 100 pp.

Krabbenhoft, D.P., C.J. Bowser, M.P. Anderson, and J.W. Valley. 1990. Estimating groundwater exchange with lakes, 1. The stable isotope mass balance method. *Water Resour. Res.* 26:2445–2453.

Majoube, M. 1971. Fractionnement en oxygene-18 et en deuterium entre l'eau et sa vapeur. *Journal of Chemical Physics* 197:1423–1436.

Rosenberry, D.O., A.M. Sturrock, and T.C. Winter. 1993. Evaluation of the energy budget method of determining evaporation at Williams Lake, Minnesota, using alternative instrumentation and study approaches. *Water Resour. Res.* 29:2473–2483.

Rosenberry, D.O., T.C. Winter, D.C. Buso, and G.E. Likens. 2007. Comparison of 15 evaporation methods applied to a small mountain lake in the northeastern USA. *Journal of Hydrology* 340:149–166.

Schulze-Makuch, D., D.A. Carlson, D.S. Cherkauer, and P. Malik. 1999. Scale dependency of hydraulic conductivity in heterogeneous media. *Ground Water* 37:904–919.

Stewart, R.B., and W.R. Rouse. 1976. A simple equation for determining the evaporation from shallow lakes and ponds. *Water Resour. Res.* 12:623–628.

Tiedeman, C.R., D.J. Goode, and P.A. Hsieh. 1997. Numerical simulation of ground-water flow through glacial deposits and crystalline bedrock in the Mirror Lake area, Grafton County, New Hampshire. U.S. Geological Survey Professional Paper 1572, 50 pp.

Winter, T.C. 1981. Uncertainties in estimating the water balance of lakes. *Water Resources Bulletin* 17:82–115.

5

EVALUATION OF METHODS AND UNCERTAINTIES IN THE CHEMICAL BUDGETS

JAMES W. LABAUGH, DONALD C. BUSO, AND GENE E. LIKENS

Comparison of chemical mass input and output associated with hydrologic processes alone is not a complete mass balance; it merely sets the stage for developing hypotheses concerning changes that might be expected to occur in storage or biological and gaseous fluxes, based on the comparison. Such hypotheses are tested by comparing the difference between mass input and output to changes in storage measured independently. A complete mass balance, also termed net ecosystem flux (Likens et al. 2002, and see chapter 1), includes change in storage, non-hydrologic sinks, and net gaseous flux, in addition to chemical mass input and output resulting from the movement of water into and out of the ecosystem. Yet, for this book, it is recognized that uncertainty also needs to be factored into any of these attempts at mass balance. In chapter 3, the balance between chemical fluxes resulting from water movement into and out of the lake was examined in relation to change in storage that was measured independently. In this chapter, we examine the uncertainty inherent in the water and chemical budgets of the lake in relation to differences between chemical mass change in storage and to the balance between solute inputs and solute outputs associated with water fluxes.

Uncertainties in the determination of water fluxes are one of three sources of uncertainty when calculating chemical mass budgets of aquatic systems (LaBaugh 1985). The other two sources of error or uncertainty are those associated with chemical analyses done to determine concentrations of chemicals in water and those associated with the way in which water samples were obtained for chemical analyses of water from atmospheric deposition, stream water, ground water, and lake water. Uncertainty in the water budget of the lake is presented in detail in chapter 4. The study of Mirror Lake in general was not designed to enable quantitative assessment of uncertainty associated with chemical analyses or sample collection, or the effects of those uncertainties in the determination of chemical budgets of the lake. During a period of 40 days (March 22 to April 30, 1993), however, samples were collected from the Northwest Tributary of Mirror Lake on a daily basis to examine uncertainty in the use of weekly samples to estimate chemical mass input from the stream (Bukaveckas et al. 1998). The long-term data from the Hubbard Brook Experimental Forest also provide useful information on the standard deviation in mass fluxes encountered in nearby watersheds (Likens and Bormann 1995). Uncertainty in water fluxes has been an integral component of the study of Mirror Lake and nearby watersheds (Winter et al. 2003; Bailey et al. 2003; Rosenberry et al. 2007). Another example of how uncertainty in mass of a chemical (phosphorus) can be related to the frequency of sample collection in combination with various methods used to determine streamflow is found in Scheider et al. (1979).

The methods used to determine the chemical budget values for Mirror Lake are presented in chapter 3. In this chapter we present more detail about considerations that went into use of chemical data to determine chemical balances and budgets of the lake, and the uncertainties in those budgets. It is useful, however, to begin with a general overview of the relation of uncertainty in the water budgets to chemical budgets.

UNCERTAINTY IN THE WATER BUDGETS USED TO DETERMINE CHEMICAL BUDGETS

In the case of Mirror Lake, no term in the water budget was calculated as the difference (residual) between water gains, losses, and change in storage. Change in storage was measured independently from atmospheric precipitation, evaporation, stream discharge into the lake, the surface water outlet of the lake, groundwater flow into the lake, and loss of water from the lake by flow into ground water (chapter 2). Thus, any imbalance (residual) in the independently measured water budget components may be compared to the cumulative error.

As noted in chapter 2, the equation (Equation 2-3) for a first-order error determination, assuming that errors are independent and randomly distributed, is

$$\delta = \sqrt{\delta_P^2 + \delta_E^2 + \delta_{SWI}^2 + \delta_{SWO}^2 + \delta_{GWI}^2 + \delta_{GWO}^2 + \delta_{\Delta L}^2}$$

Using Equation 2-3 above, the first-order error determination for the Mirror Lake water budget was calculated two ways. Prior to June 1, 1990, the percent error in surface water flow from the lake was assumed to be 15 percent (chapter 4):

(1) cumulative error $= \sqrt{\begin{array}{l}((P*0.05)^2 + (E*0.15)^2 + (SWI*0.05)^2) \\ + (SWO*0.15)^2) + (GWI*0.25)^2 \\ + (GWO*0.25)^2 + (\Delta L*0.10)^2)\end{array}}$

After June 1, 1990, the percent error in surface water flow from the lake was assumed to be 5 percent (chapter 4):

(2) cumulative error $= \sqrt{\begin{array}{l}((P*0.05)^2 + (E*0.15)^2 + (SWI*0.05)^2) \\ + (SWO*0.05)^2) + (GWI*0.25)^2 \\ + (GWO*0.25)^2 + (\Delta L*0.10)^2)\end{array}}$

where

P is precipitation,
E is evaporation,
SWI is surface water flow to the lake,
SWO is surface water flow from the lake,
GWI is groundwater discharge to the lake,
GWO is loss of lake water to ground water, and
ΔL is change in volume of water contained in the lake, in m^3 for the year.

The errors associated with measurement of each of the other components were noted in chapter 4:

precipitation, 5 percent
evaporation, 15 percent
surface water flow into the lake, 5 percent
groundwater flow into the lake, 25 percent
lake flow into ground water, 25 percent
change in lake volume, 10 percent

Cumulative error for the annual water budgets of Mirror Lake during the period 1981 to 2000 was greater than the volume represented by the residual of the water balance for 16 years, but not for the years 1983, 1984, 1986, and 1990 (Table 5-1). For most years, therefore, the imbalance in the water budget for the lake likely reflects the cumulative uncertainties in the measurement of the individual budget components, rather than indicating that some substantial source or loss of water moving into or out of the lake was not measured.

To examine the contribution of uncertainty of individual water budget components to the total, or cumulative error of the entire water budget, we assumed the error term for individual budget components was approximated by multiplying the volume of the component by the percent error of the component and squaring the result. This variance of the error of individual measured components (δ_i^2) hereafter is referred to as the first order error value or uncertainty in the individual budget component. The contribution of an individual budget component to the cumulative error was assumed to be represented by determining the percent of the variance of the total, or cumulative error (δ^2) represented by the first-order error value of that component (δ_i^2)(Table 5-2). The greatest contribution to the cumulative error in the water budget was from uncertainty in the determination of groundwater flows into and out of the lake. On average over the course of a year, loss of water from the lake by flow into ground water represented 25 percent of the entire water budget (chapter 2). Uncertainty in the determination of the loss of water from the lake by flow to ground water contributed between 48 and 89 percent of the cumulative error (Table 5-2). On average, groundwater flow into the lake represented 8 percent of the entire water budget (chapter 2). Uncertainty in the determination of groundwater flow into the lake represented between 6 and 11 percent of the cumulative error. Together, uncertainty in groundwater fluxes constituted 54 to 95 percent of the cumulative error in the water balance. The large contribution of uncertainty in groundwater fluxes to and from the lake to cumulative error depends on both the magnitude of groundwater contributions to the entire water balance of the lake, 33 percent of the total based on the annual average of 20 years, and the 25 percent uncertainty estimated for the determination of those groundwater fluxes. The assignment of 25 percent uncertainty in the determination of groundwater fluxes to and from the lake is a function of the different factors that account for that uncertainty—determination of the hydraulic gradient, cross-sectional area of flow, and hydraulic conductivity (chapter 4). In the case of Mirror Lake, additional information, stable isotope budgets and groundwater flow modeling, provides evidence that the groundwater fluxes are reasonable estimates. The general agreement between those flux values reduces the overall uncertainty in what the groundwater fluxes were for the lake. In addition, the annual chemical budget for magnesium, which in Mirror Lake is chemically conservative, is balanced based on a 20-year average (chapter 3). The balance for the annual magnesium budget is another indication that the water budgets for the period 1981 to 2000 likely are a good account of the water flowing into and out of the lake, including the groundwater fluxes, as well as change in storage.

UNCERTAINTIES IN CHEMICAL ANALYSES

In chapter 3 (Table 3-1), we provide a summary of the analytical protocols used in the determination of chemical concentrations in water collected from precipitation, streams, the lake surface outlet, ground water, and the lake. The precision over the range of concentrations measured in those waters commonly was plus or minus 5 percent. Exceptions were ammonium, plus or minus 10 percent, and dissolved organic carbon, plus or minus 2 percent.

UNCERTAINTIES IN SAMPLE COLLECTION

Advances in instrumentation have enabled us to measure some of the water budget components on a continuous basis, such as evaporation, lake level and surface water, streamflow into the lake, and surface flow out of the lake at the outlet. Atmospheric precipitation was measured on an event basis (chapter 2). Continuous measurement of the chemical characteristics of water in the tributaries flowing into the lake, the lake, ground water, and the surface outlet of the lake was not feasible during our study. Measurement of chemical characteristics of atmospheric deposition as it occurred and the relation of such measurement to sample collection on an event or weekly basis has been done previously at the Hubbard Brook Experimental Forest (Cogbill et al. 1984). Some uncertainty exists in the determination of chemical budgets as a result of the way in which chemical data obtained from samples collected at a specific point in time were used in combination with water volume data for a longer period in order to calculate change in chemical mass over time. The calculations of chemical fluxes using chemical and hydrological data in the Hubbard Brook Ecosystem Study, and associated quality assurance procedures, are described in Buso et al. (2000).

PRECIPITATION SAMPLE COLLECTION

Variability in precipitation volume between standard rain gauges at Hubbard Brook Experimental Forest is on the order of 10 percent (Bailey et al. 2003), but the chemistry of bulk collectors arrayed across the Hubbard Brook Valley is nearly identical on a weekly basis (Likens et al. 2002). Based on these data, we determined that an additional chemical collector at Mirror Lake was unnecessary. However, year-round, weekly collections of precipitation are necessary, since temporal variability in chemical characteristics of precipitation can be several orders of magnitude greater than spatial variability at the Hubbard Brook Experimental Forest (Buso et al. 2000). Likens (1985) indicated in general that concentrations in atmospheric precipitation were not correlated with elevation, except for sulfate and nitrate. The collector for chemical analyses of atmospheric deposition for the Mirror Lake basin is located at an elevation of 252 m, whereas the elevation of the lake is 213 m. Chemical analyses of atmospheric deposition from the collector in the watershed of the lake at the elevation of 252 m were assumed to represent what was falling on the lake, based on the findings of Likens (1985).

SURFACE WATER SAMPLE COLLECTION

Tributary Inflow
In the case of Mirror Lake, daily streamwater chemical inputs via the three tributaries were calculated by multiplying the weekly chemical analyses and the daily streamwater volumes to produce a daily flux for each discreet sample date. For days between discrete samples, when no actual streamwater sample was collected, the average concentration determined from the previous and next streamwater sample was applied to the daily water volume to create a daily flux value. These calculations of mass per unit time were consistent with the way in which calculations have been done in the nearby Hubbard Brook Experimental Forest watersheds for the past 45 years (Likens and Bormann 1995; Buso et al. 2000).

The determination of mean absolute error in mass flux for the tributaries flowing into Mirror Lake using the method of calculation described above may be accomplished when continuous chemical analyses can be related to continuous discharge measurement. Bukaveckas et al. (1998), however, used a short period (22 March to 30 April, 1993) of daily water sample collection from the Northwest Tributary to Mirror Lake to examine uncertainties in chemical mass flux determinations. Based on that 40-day period, use of average concentrations between weekly samples resulted in mass flux for the period that was within 5 percent of the mass flux determined from measured daily concentrations and flow for calcium, sodium, and sulfate, and within 10 percent for chloride and dissolved silica. Also, Stelzer and Likens (2006) indicate that sample collection frequency for dissolved silica in moderately flashy streams needs to be weekly to keep the calculated mass flux within about 7 percent of the mass calculated from daily sample collection.

One of the few studies to document uncertainty associated with a variety of methods to calculate chemical mass flux of streams flowing into a lake was done by Scheider et al. (1979). Scheider et al. (1979) indicated that calculation of mass by use of continuous discharge measurements with average concentration between times when discrete samples were collected for chemical analysis (as done herein for Mirror Lake) has a mean absolute error of 3 percent, in comparison with the results provided by what Scheider et al. defined as the best method. In the Scheider et al. study, the best method was obtained by multiplying the concentration from the discrete sample by continuous discharge, where the date of sample collection defined the midpoint of the time interval of the discharge record used in the calculation. The Scheider et al. (1979) calculation of mass was not dependent on having concentration correlated with flow. Using the Bukaveckas et al. (1988) 40-day study of Mirror Lake and the Scheider et al. (1979) examination of uncertainties associated with various methods of mass flux calculations as a guide, it is expected that the uncertainty introduced by the method of calculation of mass input from the tributaries to Mirror Lake is about 3 to 5 percent.

Surface Outflow

During low-flow conditions, samples taken from below the spillway on the bedrock lobe on which the dam is placed have higher pH, ANC, and DIC values than those collected from a few meters above on the lake spillway itself. Calcium and magnesium concentrations were also larger in samples taken from below the spillway. The lower site may incorporate rock fracture flow that is not part of the lake hydrologic balance, or some deep seepage that has been chemically altered after leaving the lake through the sediments. For the purposes of monitoring surface outflow from Mirror Lake, the upper spillway site was selected as the better choice for measuring actual lake outflow chemical characteristics. Calculation of mass flux out of the lake was done in the same way as for stream input—daily chemical loss via the surface outlet was calculated by multiplying the weekly chemical analyses and the daily surface outlet volumes to produce a daily flux for each discrete sample date. For days between discrete samples, when no lake outlet sample was collected at the dam, the average chemistry of the previous and next surface outlet samples was applied to the daily water volume to create a daily flux value. Owing to the fact that the method of combining continuous flow data and discrete chemical data to determine the surface outlet mass flux was similar to that used for determining surface tributary mass flux, the mean absolute error is likely similar, about 3 to 5 percent. This expectation is consistent with the fact that Bukaveckas et al. (1998) found that chemical mass loss from the lake through the surface outlet determined using average concentrations derived from weekly chemical analyses was within 2 percent of the chemical mass determined from daily chemical analyses during a 40-day period.

Lake Water
Samples from the lake were commonly collected at 0.5m, 2m, 4m, 6m, 8m, and 10m below the water surface, although this vertical sequence varied more prior to 1990, when collections were also made by colleagues. The maximum water depth of Mirror Lake during the study was 11 m (Likens 1985). Although theoretically it would be more accurate to measure from a fixed point on the bottom toward the lake surface, using the lake's surface as a reference does not introduce more than about 0.4 m error in sampling depth, because the lake level is relatively constant in this wet climate (see chapter 2). Furthermore, volume-weighting the lake chemistry based on six separate sample points reduces bias produced by small errors in depth.

Lake Ice Effects on Annual Storage Estimates One phenomenon intrinsic to water and chemical budgets in lakes within the northern Temperate Zone is the effect of ice cover on quantifying nutrient pools and inputs. In late March or early April after cold winters with a thick (>0.5 m) snow and ice cover, the uppermost lakewater sample may have been collected from as much as 1 m below the surface. Under such circumstances, the ice cover on Mirror Lake can approach about 15 percent of the total lake volume, and this solid volume might need to be compensated for in calculations of lake nutrient pools. Initially, lake ice tends to exclude solutes as it forms, and when this clear ice thaws, it can be highly dilute. Simultaneously, this surface also can incorporate frozen precipitation inputs (rain and snow), and particulate material, which can accumulate over the dilute lake ice and may not enter the lake until months after deposition. With the exception of ammonium, hydrogen ion, nitrate, and phosphate, the precipitation inputs for most solutes are relatively small compared to the streamwater or groundwater annual inputs (chapter 3).

Atmospheric deposition on top of lake ice begs the following question: when does atmospheric deposition actually become an input, and when does it effect the organisms and environment of the lake? This complex mixture of dilute and concentrated material creates a difficult problem in calculating water and chemical budgets because it involves combining pool changes and inputs across an ecosystem boundary, often with a lengthy temporal delay. Our estimates of annual lake solute pools were made on 31 December, which was a date when the lake ice averaged only 0.15 m thick. This thickness represented 2.4 percent of the total lake volume (862,000 m^3; Likens 1985) on average, and only one month of potential (i.e., frozen) atmospheric deposition at most, given an average ice-in date of 6 December (Likens 2000). We concluded that the impact of this thin early ice stratum on the whole-lake solute pool was insignificant compared to the other sources of uncertainty, such as bathymetric error or even analytical precision. Thus, we have ignored the chemical characteristics of the ice in our annual calculations and added the frozen precipitation on the ice into the annual budget during the month in which it fell (chapter 3).

Lake Flow into Ground Water

As noted in chapter 3, water flowing from Mirror Lake into ground water primarily occurs in the 0 to 2 m depth range of the lake. The concentrations in weekly samples collected at the surface outlet were similar to concentrations in the four to six samples collected from the 0 to 2 m depth interval in the lake each year. The out-seepage flow data were compiled on a monthly basis (chapter 2), so the volume-weighted monthly average surface outlet concentration data were multiplied by the measured monthly out-seepage flow to calculate monthly solute outflux. Uncertainties in this calculation are not easy to quantify. It is possible that the lake was not well mixed throughout the 0 to 2 m depth across the lake, so that concentrations at the surface outlet might not adequately represent lake water flowing into ground water. It also is possible that streamflow into the lake on occasion might flow along the shoreline such that flow out of the lake along that shoreline would have concentrations representative of one of the streams, rather than of the lake. Based on the 20 years of observation, however, it is most probable that the surface outlet is an adequate indicator of concentrations in lake water that flows from the lake into ground water.

GROUNDWATER SAMPLE COLLECTION

Groundwater Flow into Lake
Other multi-year studies of the chemical characteristics of ground water in the watersheds of lakes provide a useful context for examination of groundwater sample collection at Mirror Lake and use of resulting chemical data in mass flux determination in chemical budgets (LaBaugh 1991; Krabbenhoft and Webster 1995; Wentz et al. 1995). Such studies indicate that spatial variation in chemical characteristics between wells commonly is greater than temporal variation in chemical characteristics in an individual well. Thus, in some studies, chemical characteristics of wells representative of different areas of groundwater flow into a lake, called "index" wells, are used with groundwater flow from the areas the wells represent to calculate groundwater chemical mass input to a lake (LaBaugh et al. 1995, 1997). The substantial variability that may exist in chemical characteristics of ground water can be a function of position of the sampled ground water within the groundwater flow system (Hem 1989). Smaller concentrations tend to be found in areas where ground water readily receives recharge and the water has not had much time to interact and react with the geological material through which it moves. Larger concentrations tend to be found some distance along a flow path, whereby ground water has had some time to interact and react with the surrounding geologic material. Ground water and pore waters in the immediate vicinity of the shoreline of a lake can vary substantially in chemical characteristics over time due to flow reversals (Krabbenhoft and Webster 1995; Wentz et al. 1995; Rosenberry et al. 1999).

Calculation of the groundwater contribution to the solute budgets of Mirror Lake used methods similar to those found in published literature and described above. Ground water was collected from many locations within the watershed of Mirror Lake. Regular and repeated collection of ground water for chemical analysis over many years, however, was not part of the data collection procedures for the study of Mirror Lake. As noted in chapter 3, chemical characteristics of ground water obtained from the glacial till indicate substantial spatial variability, a trait consistent with similar studies of ground water–lake interaction. Smaller concentrations were found in areas of the watershed where ground water was recharged. Some of the wells within a few meters of the shoreline were influenced by movement of lake water into ground water even though the overall gradient was that of groundwater flow into the lake. For the purpose of determining groundwater flow into Mirror Lake, index wells were used to calculate flow entering the lake from different segments of the shoreline of the lake (chapter 4). The wells considered to represent different areas of the groundwater flow field flowing into the lake were not always those from which chemical characteristics of ground water were available. Thus, it was decided to approximate the chemical characteristics of ground water flowing into the lake by using the median concentration of all wells for which chemical analyses were obtained from samples analyzed at the Robert S. Pierce Ecosystem Laboratory. The use of a fixed median value for incoming groundwater concentrations meant that any variability in groundwater influx was due exclusively to changes in groundwater flow.

ALTERNATIVE APPROACHES TO DETERMINING CHEMICAL BUDGETS

CHEMICAL MASS INPUT FROM TRIBUTARIES

The method chosen to calculate chemical mass input from the three tributaries flowing into Mirror Lake was consistent with literature (Scheider et al. 1979) indicating the calculation is one of the available options associated with the least uncertainty. The method used, however, was not the only option regarding the way in which chemical mass input can be calculated. An alternative method of calculating chemical mass input from the tributaries is to use the results of a simple least-squares regression model between concentration and flow (Bukaveckas et al. 1998). The statistical relation between concentration on the day samples were collected and the corresponding daily flow for that day enables estimation of concentration as a function of daily flow on days for which no sample was collected.

In the case of Mirror Lake, the regression model used to examine alternative ways to calculate chemical mass input over time was [concentration] = $a*$ (Log_{10}(daily flow)) $\pm$ b_0 (Table 5-3). Note that concentrations obtained from sample collection on a particular date were used to develop the relation to daily flow. The observed relation may be refined in the future when it may be possible to relate continuous concentration data to continuous flow data within a day. Furthermore, this alternative was used only for those chemical constituents for which the relation between concentration and flow was statistically significant— that is, calcium, magnesium, sodium, hydrogen ion, chloride, and dissolved silica concentrations (Table 5-3). No relation was statistically significant between daily flow and concentration measured on the date that flow was recorded for potassium, sulfate, ANC, phosphorus, nitrate, ammonium, or dissolved organic carbon. On the basis of a 40-day period of daily sample collection for chemical analyses, Bukaveckas et al. (1998) found that the use of regression models to determine chemical flux for calcium, magnesium, sulfate, chloride, and dissolved silica from the Northwest Tributary to the lake resulted in values within 5 percent of chemical mass flux determined from daily chemical analyses. Except for sulfate, the chemical mass fluxes determined using the regression model were approximately half or less of the percent deviation from the chemical mass flux determined from daily values than was the case for chemical fluxes determined from using average concentrations derived from weekly samples.

The concentrations used to determine solute mass input from the tributaries were determined in two ways. The first method, referred to hereafter as the "average" method, used average concentration of samples collected approximately a week apart (as presented in chapter 3). The second method, referred to hereafter as the "model" method, used a regression equation model (relating concentration to discharge). The two methods of determining solute mass input produced similar values (Fig. 5-1). On the basis of 20-year average values (Table 5-4), the regression model–derived chemical mass inputs for calcium, magnesium, and dissolved silica were slightly less than chemical mass inputs derived from average concentration between sample collection dates in all three of the lake sub-watersheds. The same was true for sodium and chloride in the Northwest and West Tributaries, but not in the Northeast Tributary. The 20-year average difference between the chemical mass input determined by the two methods, expressed as a percentage of the mass determined using the average concentration between sample dates, was generally small, ranging from 0.8 to 5.6 percent. Such differences are consistent

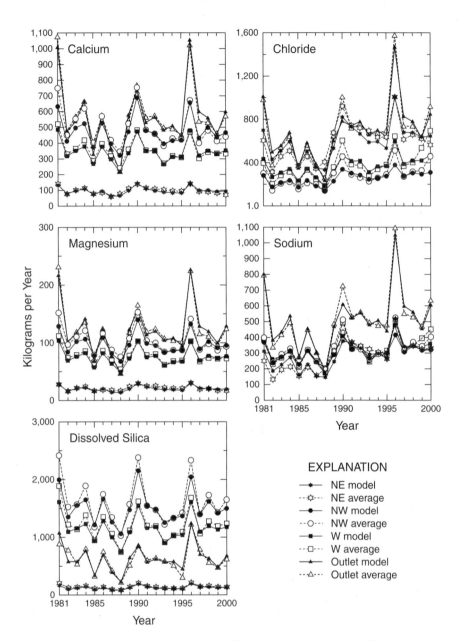

FIGURE 5-1. Annual solute mass in the tributaries and surface outlet of the lake determined using concentrations derived from the regression between flow and measured concentration (model) and measured concentrations (average). Solutes shown are those solutes for which a statistically significant relation existed between solute concentration and flow.

with the differences indicated by the analysis of Bukaveckas et al. (1998). The differences between the solute mass determined using the average concentration between sample dates and the concentration derived from the regression model commonly were small in individual years as well (Fig. 5-1). The relation between the solute mass derived from the two methods was not always the same each year. In 1981, for example, the calcium mass in the Northeast Tributary was larger based on the model than based on the average concentration, whereas in 1997, the model-derived value was less than that derived from the average.

CHEMICAL MASS LOSS BY SURFACE OUTFLOW

The method chosen to calculate chemical mass loss from Mirror Lake by the surface outlet was not the only option regarding the way in which chemical mass input can be calculated. As in the case of tributary inflow to the lake, concentrations of some chemical constituents could be related to flow, based on simple least-squares regression analysis (Table 5-3) The chemical constituents related to flow were the same as for the tributaries flowing into the lake: calcium, magnesium, sodium, hydrogen ion, chloride, and dissolved silica. In the case of the outlet, Bukaveckas et al. (1998) found the chemical mass lost from Mirror Lake through the surface outlet determined by use of regression models was within 2 to 10 percent of the mass determined from measured daily concentrations for a 40-day period in the spring of 1993. In addition, the chemical mass loss determined from the average of weekly samples was closer to the mass determined from daily measured concentrations (i.e., between 1 and 3 percent). For the 20 years presented herein, estimates of chemical mass loss from the lake through the surface outlet for calcium, magnesium, and dissolved silica were slightly larger using the regression model than those for mass determined using the average method (Table 5-4). Chemical mass loss of sodium and chloride through the surface outlet was slightly less using the regression model than for mass determined using the average method. The percent difference between the two methods was 0.4 to 5 percent based on the annual average values for the years 1981 to 2000.

Using a regression model of silica and flow, Stelzer and Likens (2006) simulated daily dissolved silica concentrations in 14 streams, including Watershed 6 of the Hubbard Brook Experimental Forest. The chemical flux in the streams based on those simulated daily concentrations was compared with chemical flux determined from concentrations determined at 7-, 14-, 28-, 56-, and 91-day intervals. Annual fluxes determined from concentrations at these intervals were found to be larger than flux determined from simulated daily concentrations. In the case of Mirror Lake, some other chemicals did have a statistically significant relation with flow. Therefore, it is useful to examine what effect using this alternative method of calculating mass flux related to surface water flow in and out of the lake has on chemical budgets of the lake. Given the small differences between the chosen method of calculation of chemical mass flux for the tributaries and the surface outlet used in chapter 3, and the alternative method shown in Table 5-4, the effect on the overall budget for Mirror Lake was small (Table 5-5).

The difference between inputs, outputs, and change in storage obtained using chemical mass inputs and outputs for surface water using average concentration between chemical sample collection dates and using concentration determined from the regression relation between concentration and flow were not statistically significant (Table 5-6). The residual for each of the two methods used to determine surface water contributions to the overall budget was tested to determine whether the residual was not representative of a normal distribution using the Shapiro-Wilk test at $\alpha = 0.1$. For those constituents for which residuals obtained from both methods were normally distributed—calcium, magnesium, sodium, and chloride—comparison of the residuals between the two methods was done by t-test. For dissolved silica, the residual obtained from use of the chemical mass determined from concentration estimated from the regression model was not normally distributed, so comparisons were made using a non-parametric median test (Steel and Torrie 1980).

Chemical budgets can be used to evaluate whether a lake is expected to retain or lose chemicals, based on the ratio of chemical outputs to chemical inputs (chapter 3). Chemical balance data presented in chapter 3 indicate that Mirror Lake retains a substantial amount of the input of dissolved silica, but does not retain substantial amounts of calcium, magnesium, or sodium, and loses chloride. Recall that these constituents are the only ones besides hydrogen ion that had a significant relation with flow. The use of a regression model to determine concentrations to multiply by flow to obtain chemical mass input or output from surface water does not change the general relation between chemical inputs and outputs for dissolved silica, calcium, magnesium, or sodium with respect to the retention values presented in chapter 3 (see Table 5-7). The absence of a statistically significant relation between flow and dissolved inorganic nitrogen, phosphate, potassium, or sulfate concentrations is why a similar comparison for those chemicals is not presented in Table 5-7.

CHEMICAL MASS INPUT FROM GROUND WATER

As noted herein, the median concentrations of chemicals found in groundwater samples collected in the watershed of Mirror Lake were used to determine chemical mass input from ground water to the lake. Owing to the fact that flow in the tributary streams is maintained by ground water during minimal flow in those streams, concentrations in the Northwest and West Tributaries during minimal flows might provide an indication of concentrations in ground water that reaches the lake. Concentrations in the Northeast Tributary are affected by salt applied to the nearby highway, I-93; therefore, data from that tributary were not used to approximate median concentrations of ground water flowing into the lake. The possibility that concentrations in the inflowing streams might approximate groundwater flow into the lake was examined in two ways. In the first case, when flow in the Northwest and West Tributaries was 1 percent or less of maximum flow for the 20-year record, concentrations from samples collected at that time were used to determine the median concentration used in the calculation of chemical mass input from ground water (Table 5-8). In the second case, when flow was small and observed concentrations were relatively constant, those constant concentrations at low flow in the Northwest and West Tributaries were used to determine the median concentration used in the calculation of chemical mass input from ground water.

For Mirror Lake, no matter which concentration was used to represent groundwater flow into the lake, the concentration was multiplied by groundwater flow into the lake to determine chemical mass input to the lake from ground water. A constant concentration was used for all 20 years. Thus, rather than compare the solute mass obtained using different concentrations to represent ground water, comparisons are made based on concentrations. The relation between the median concentration in ground water and the median concentration in the Northwest and West

Tributaries at low flow was not consistent for all solutes. In the case of chloride, phosphate, and DOC, median values in the tributaries at low flow were nearly double or more than the median value in ground water. In contrast, at low flow, median values in the tributaries for potassium, DIC, and ANC were approximately half or less than in ground water. The median value of dissolved silica was somewhat less in ground water than in the two tributaries. Nitrate was larger at low flow at constant concentration than at low flow or in ground water, based on median values. For the remainder of the solutes, the differences between values were less pronounced. Median concentrations of groundwater samples were somewhat larger than concentrations from the tributaries at low flow for calcium (~18 percent larger), sodium (8 percent larger), hydrogen ion (10 percent larger), and sulfate (22 percent larger). In the case of magnesium and ammonium, the median value in ground water was the same as in the two tributaries at low flow when concentration was constant, but not identical to values at low flow. These data indicate that the concentrations in the tributaries are similar to ground water in general, but are not exact representations of median concentrations in ground water.

To examine the effect of using concentrations in the two tributaries at low flow as a proxy for ground water, it is useful to examine the resulting chemical budgets of the lake with respect to chemical retention (Fig. 5-2). On the basis of a comparison of chemical inputs and outputs derived from hydrologic inputs and outputs, the lake retained hydrogen ion, dissolved silica, inorganic nitrogen, and phosphate (inputs exceeded outputs) in each of the 20 years of study. Note that this comparison of inputs and outputs does not account for particulate matter input or output, gaseous exchange, or biotic transfer. Except for phosphate, the percent retention was commonly more than 50 percent of input to the lake for these chemical elements. The percent retention of phosphate generally increased during the 20 years of study, and percent retention of ammonium appeared to increase as well. In contrast, retention of calcium, magnesium, sodium, potassium, sulfate, and chloride were consistently less than 30 percent of input to the lake, except for sulfate in 1990. In many years, this same group of chemicals also had negative values of retention, and there was no obvious trend with time. Overall, retention was approximately the same for chemicals in Table 5-8 that had slight differences in concentration between the different values used

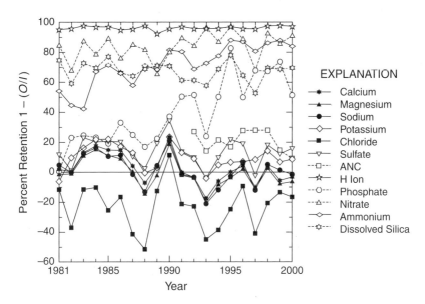

FIGURE 5-2. Chemical retention in Mirror Lake, based on the relation between all chemical inputs and outputs as described in chapter 3.

to represent ground water (Table 5-9). The overall relation between all inputs and outputs for each chemical (as percent retention in the lake) to that presented in Fig. 5-2 are similar when using concentrations in the West and Northwest Tributaries at low flow to approximate groundwater concentrations in incoming ground water (Fig. 5-3). For phosphate, the proxy tributary concentrations used in chemical flux calculations for ground water were more than twice the value in ground water. Thus, the retention values for phosphate resulting from the proxies were somewhat more than for ground water (Fig. 5-4a, Table 5-9). For potassium, the proxy tributary concentrations used in the chemical flux calculation for ground water were about half the concentration in ground water. Thus, the retention values for potassium resulting from use of the groundwater proxy concentrations were somewhat less than for retention obtained from using the groundwater concentration. In the case of potassium, the relation between inputs and outputs on an annual basis changed from retention in most years (using the median of all wells) to export (using tributary concentrations to approximate groundwater concentrations) (Fig. 5-4b).

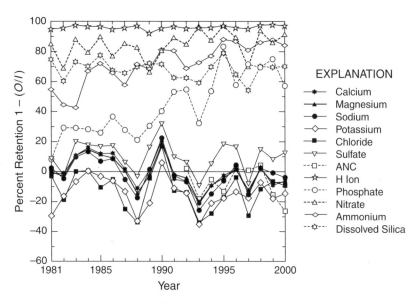

FIGURE 5-3. Chemical retention in Mirror Lake, based on the relation between all chemical inputs and outputs using median concentrations of the Northwest and West Tributaries at low flow as proxies for groundwater concentration in the calculation of chemical input from ground water.

CHEMICAL MASS LOSS BY FLOW THROUGH GROUND WATER

The solute concentrations used to determine chemical mass loss from Mirror Lake through ground water were concentrations measured in the surface outlet (chapter 3). Alternative estimates of the contribution of chemical mass loss from Mirror Lake to ground water result from assuming the concentrations of the lake profile represent lake water flowing into ground water, and assuming concentrations of the surface outlet represent lake water flowing into ground water. The first alternative represents fewer samples on an annual basis than does the second. Samples were collected from the surface outlet on a weekly basis, whereas commonly sample collection from the lake was done four to six times per year.

The 20-year volume-weighted concentrations of solutes were commonly similar among the lake, the surface outlet, and that calculated as lake water flowing into ground water (Table 3-8). Phosphate and ammonium concentrations were the same for the lake, the outlet, and lake flow to ground water (Table 3-3). The long-term volume-weighted

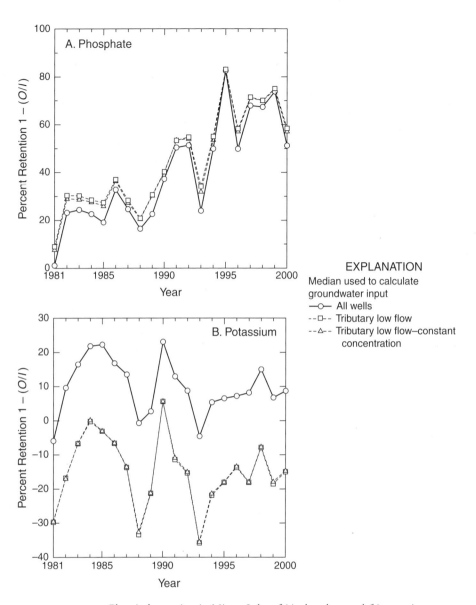

FIGURE 5-4 Chemical retention in Mirror Lake of (a) phosphate and (b) potassium, based on the relation of all chemical inputs and outputs, using three different methods to determine groundwater input (median concentration of all wells, median concentration of water from the Northwest and West Tributaries at low flow, and median concentration of tributary water at low flow when concentration was constant).

concentrations of calcium, magnesium, sodium, potassium, chloride, and sulfate in the surface outlet were approximately 2 to 6 percent less than attributed to ground water out, whereas ANC, phosphate, ammonium, and DIC concentrations were identical. Concentrations in the lake of calcium, magnesium, sodium, potassium, chloride, sulfate, and ANC were approximately 2 to 6 percent larger than attributed to lake flow to ground water. Surface outlet concentrations of dissolved silica were larger than those in the lake or those attributed to lake flow to ground water. As a result of these small differences, the differences among percent retention values resulting from using the long-term average concentrations attributed to lake flow to ground water, the surface outlet, or the lake for each solute were generally unremarkable: calcium (2.23, 4.81, −0.58), magnesium (−0.77, 2.87, −3.19), sodium (1.40, 4.56, −1.69), potassium (9.74, 13.23, 8.59), chloride (−22.86, −18.15, −26.76), sulfate (14.07, 14.89, 13.31), ANC (20.76, 20.76, 10.54), hydrogen ion (96.07, 95.73, 95.73), silica (67.22, 65.19, 66.9), phosphate (all 39.75), nitrate (83.83, 78.31, 86.04), ammonium (all 71.85), DOC (2.52, −1.92, 6.7), and DIC (36.72, 36.72, 36.48).

THE RELATION OF UNCERTAINTIES TO HYPOTHESES

Comparison of chemical inputs and outputs resulting from movement of water into and out of a lake, commonly termed retention in the limnological literature (Dillon and Rigler 1974; Schindler et al. 1976; Reckhow and Chapra 1979), as well as net hydrologic flux (Likens et al. 2002) or balance (Andersson-Calles and Eriksson 1979) in the watershed literature, provides a useful starting point for developing hypotheses about the processes controlling the movement and fate of solutes in a lake. In simple terms, change in mass storage in the lake equals inputs minus outputs (chapter 1).

$$\Delta S = I - O \qquad (5\text{-}1)$$

where

ΔS is solute mass change in storage,

I is solute mass input, and

O is solute mass loss.

In the case where solute mass input is greater than solute mass loss, the change in storage is positive; solute mass in the lake increases. In the case where solute mass input is less than solute mass loss, the change in storage is negative; solute mass in the lake decreases. In the case where solute mass input is equal to solute mass loss, the change in storage is zero. When ΔS, I, and O are each measured independently, then the imbalance between I and O can be compared to ΔS to determine if the values are indeed equivalent. In the following discussion, the mass represented by the annual imbalance between chemical inputs and outputs will be referred to as the input-output imbalance. Also, the annual solute mass change in storage measured in the lake will be referred to as change in storage.

The chemical mass represented by the input-output imbalance was not the same as the change in storage for all of the chemical constituents examined in the Mirror Lake study (Table 3-9, chapter 3, Fig. 5-5) because as noted in chapter 3, the change in storage was determined by comparing solute mass in the lake at the beginning of the year with the solute mass in the lake at the end of a year. Solute mass in the lake was determined from volume-weighted average concentrations in the lake obtained from samples collected within the lake four to six times per year. A plot of the volume-weighted averages was used to interpolate the volume-weighted solute concentration at the beginning and end of a year. The values of the input-output imbalance were approximately the same (within 2 percent) as the change in storage only during a few years of the study. Examples of such rare occurrences were 1982, 1991, and 1992 for calcium; 1982, 1987, 1992, 1995, and 1999 for magnesium; 1984 for sodium; 1987, 1994, 1995, 1997, and 2000 for potassium; 1981 and 1996 for chloride; and 1982, 1993, and 1997 for sulfate. In 1993, the values were within 2.5 percent ANC. The apparent similarity in 1982 for ammonium in Fig. 5-5.2 is a function of the scale of the y axis; the difference between values of chemical mass obtained by differences between input and output and the values of measured change in mass in the lake was 22 percent. Owing to the uncertainties in determining the volumes of water input and output, as well as lake volume, it would be remarkable if these values were identical or nearly so in each of the 20 years presented herein.

It is useful, therefore, to examine the input-output imbalance in relation to the change in mass in storage in the context of the absolute value

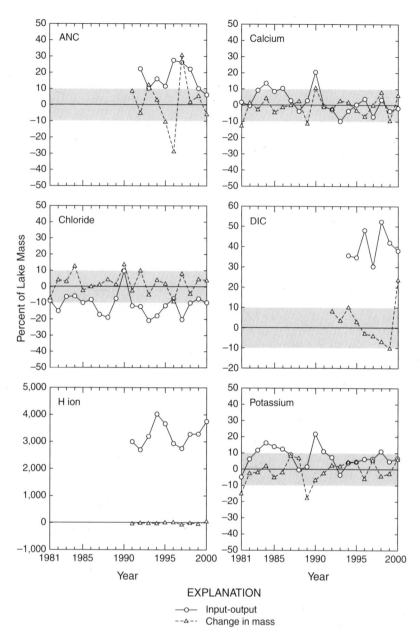

FIGURE 5-5.1. The solute balance (hydrologic flux solute mass input minus solute mass output, I–O) and change in solute mass in the lake (dmass) expressed as percent of the total mass in Mirror Lake, 1981 to 2000. Shaded band represents the bounds of ± 10% of lake mass.

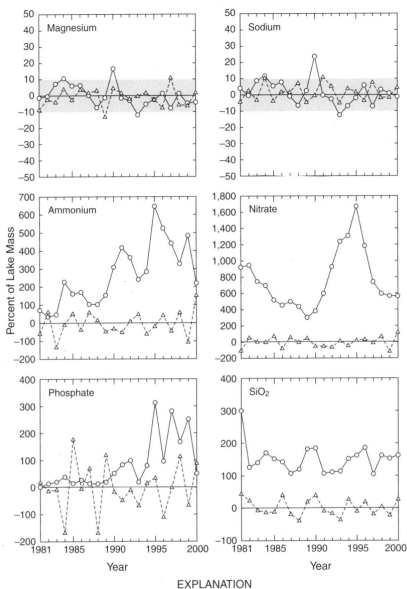

FIGURE 5-5.2.

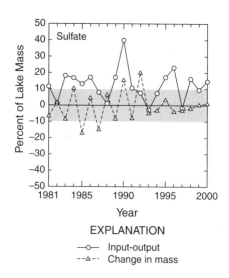

FIGURE 5-5.3.

of mass in the lake. The likely uncertainty of the volume of the lake of ± 5 percent or ± 10 percent is used here to approximate the uncertainty in the change in mass in the lake. Results of these comparisons indicate that calcium, magnesium, sodium, potassium, chloride, sulfate, and ANC generally were within the ± 10 percent band (Fig. 5-5). A noticeable disparity existed between the expected change in mass based on the input-output imbalance and the change in storage for hydrogen ion, phosphate, nitrate, ammonium, and dissolved silica. The disparity is consistent with the fact that it is likely that processes within the lake, particularly biological processes in the case of phosphate, nitrate, ammonium, and dissolved silica, remove these solutes from the water mass of the lake (see chapters 1, 3, and 6). Measurement of total phosphorus, total dissolved nitrogen, total nitrogen, and total organic carbon may reveal the importance of processes within the lake affecting changes in the dissolved phases of those solutes in future studies, but it is not considered here.

When independently determined ΔS is not equal to measured I minus measured O, the discrepancy raises the possibility that some flux or process has not been accounted for in one or more of the terms of Equation 5-1. The focus of this book has been on the determination of solute fluxes and change in storage that accompany changes in water fluxes and water mass. Yet we recognize that other factors such as particulate flux, gas exchange, sedimentation, and biotic activity will have an effect on solute input to the lake, solute loss from the lake, or change in solute storage within the lake.

In many studies of lakes, ground water has been overlooked, or determined as the volume resulting from subtracting measured water losses from measured water inputs and change in storage (see Winter 1981). Owing to the fact that either ground water or some other water source or water loss may not be accounted for, both sides of Equation 5-1 might not be equal. For Mirror Lake, however, the 20-year study was done with the purpose of including the best available techniques to quantify groundwater fluxes to and from the lake. The various techniques used to examine groundwater fluxes presented in chapters 2, 3, and 4 provide evidence that the groundwater fluxes have been well characterized. Thus, as noted in chapter 4 and elsewhere (Healy et al. 2007), initial estimates of groundwater fluxes based on a single method were revised to reflect the concurrence of the estimate determined from multiple methods. Consequently, when both sides of Equation 5-1 are not equal, the discrepancy may be due to some unmeasured process, such as particulate input, gas exchange, sedimentation, or biological sequestration and loss. Furthermore, the discrepancy may simply be a result of the uncertainty inherent in the measurement of solute fluxes and change in storage associated with hydrologic fluxes. Taking into account unmeasured processes and uncertainty in measured fluxes and storage, Equation 5-1 can be rewritten as follows:

$$R = I - O - \Delta S \qquad (5\text{-}2)$$

where

R is the residual of the solute balance.

The residual includes both unmeasured fluxes and processes and uncertainty in measurement. Independent measurements of fluxes and processes unmeasured in a water balance accounting can provide further refinement of the solute budget of the lake, leaving only uncertainty. The solute budget, as determined herein for Mirror Lake, does provide the opportunity to relate cumulative error—a measure of the solute budget uncertainty—to the residual of the solute balance.

SOLUTE BUDGET RESIDUALS

The residual of the solute budgets for Mirror Lake commonly was a value other than zero (Table 5-10). In fact, the residual value of each solute budget was seldom zero, but tended to fluctuate between positive and negative values on an annual basis. What accounts for this? At first glance, we might consider that some solute input or output associated with water fluxes has been underestimated, overestimated, or missed altogether, such as particulate input from the atmosphere (e.g., Cole et al. 1990). It also is possible that some process is occurring in the lake that either adds or subtracts solute from the lake water mass, a process not accounted for in the measurement of the water budget and associated solute budgets, such as gas exchange or sedimentation. In the absence of consideration of uncertainty in measurement of solute fluxes associated with water fluxes, further investigation might pursue uncovering previously unaccounted for sources or losses of solute from the lake. Yet another possibility exists. Even in a study such as that of Mirror Lake, which has used the best available methods to quantify water and solute fluxes, these methods still may have sufficient uncertainty such that the residual of the solute budget is indistinguishable from the uncertainty inherent in the measurement of fluxes and change in storage. In spite of this uncertainty, further studies of processes that may affect the solute budget of the lake may provide additional understanding of the controls on solute concentrations in the lake and how they change over time.

CUMULATIVE ERROR IN SOLUTE BUDGETS

In the case of Mirror Lake, the residual of the solute budget for some solutes is indistinguishable from the cumulative error in the measured components of the solute budget (Table 5-10). The absolute value of the residual of the solute budget for calcium, magnesium, sodium, potassium, and DOC were commonly smaller than the cumulative error in the solute budget for those solutes, with some exceptions. The cumulative error was larger than the residual for 14 years for calcium, 17 years for magnesium, 13 years for sodium, 13 years for potassium, and 3 of 4 years for DOC. Thus, for these solutes, when the residual is less than the cumulative error, further interpretation of the residual as representing a pattern or process, or as being indicative of some overlooked phenomenon within the lake or its watersheds, is not tractable because the residual is indistinguishable from uncertainty in the solute budget.

From 1981 to 2000, the absolute value of the residual of the solute budget was always larger than the cumulative error of the measured components of the solute budget for hydrogen ion, phosphate, nitrate, ammonium, and dissolved silica (Table 5-10). For these solutes, further interpretation of the residual as representing a pattern or process, or as being indicative of some overlooked phenomenon within the lake or its watersheds, is useful. Given the fact that the water budget of the lake was measured with the best available methods, it is most likely that particulate input (e.g., Cole et al. 1990) or biotic uptake or flocculation followed by sedimentation of these solutes may be affecting the solute budget of the lake. In the case of nitrogen, gas exchange may play a role as well.

Unlike the other solutes studied herein, sulfate, chloride, ANC, and DIC data showed no consistent relation between the residual of their solute budgets and the cumulative error in their solute budgets (Table 5-10). In 10 of the 20 years for sulfate, in 11 of the 20 years for chloride, and in 6 (1994–1999) of the 7 years for DIC, the absolute value of the residual was larger than the cumulative error in the solute budget. In 5 of the 9 years for which data were available, the cumulative error of the solute budget was larger than the residual of the solute budget for ANC. The difference between the values, however, may indicate that the residual represents some unaccounted for intermittent flux or process that may be occurring, affecting the solute budget for sulfate, chloride, ANC, and DIC. The 20-year average value of the absolute value of residual of the solute budget for sulfate (494 kg yr^{-1}) was more similar to the average value of the cumulative error in the solute budget (463 kg yr^{-1}) than for chloride (348 to 272 kg yr^{-1}), ANC (12,757 to 11,256 eq yr^{-1}), or DIC (47,270 to 23,539 mol yr^{-1}).

It is useful at this point to examine what processes might account for the mass of solute in the residual that exceeds the cumulative error, on average. In the case of chloride, on the basis of the 20-year average, the lake loses 76 kg per year. For reference, the mass in the lake was between 1989 and 3111 kg. Therefore, the average loss of chloride represents between 3.8 and 2.4 percent of the mass in the lake in a year. If we examine just those 11 years in which the residual exceeds the cumulative error for the chloride budget, then the residual exceeds the cumulative error by 217 kg per year, on average. Such losses would be expected to be associated with a decline in lake concentration. Yet in Mirror Lake, chloride concentrations have been increasing between 1981 and 2000. What might account for this? Assuming we indeed have measured the water budget well, and that there is minimal bias in chemical analysis for chloride, there likely is some other source of chloride that is unaccounted for. A likely source is from ground water. The groundwater contribution to the lake used the median concentrations of all wells. It is possible that although groundwater flow into the lake is diffuse, there may be areas of focused discharge that may have larger concentrations than those common throughout the watershed of the lake. If focused discharge occurred in the vicinity of the Northeast Tributary, for example, then the larger values of chloride in that watershed might account for the chloride that is then exported from the lake. We would expect, however, that the source of chloride would be associated with a cation, such as sodium, for example. The relation between the residual and the cumulative error for the annual sodium budget was not identical to chloride in all years, and in many years the residual was indistinguishable from the cumulative error. Yet, in two years of the study, 1988 and 1997 (Table 5-10), the residual was larger than the cumulative error for both sodium and chloride. In those years, the budget indicates that part of the residual in the annual budget for sodium and chloride is distinguishable from cumulative error. In both years, the residual for both solutes was negative, consistent with an association in the source of both solutes, heretofore unaccounted for in the measured water and solute budgets. In terms of equivalents, however, the amount by which the residual of the chloride budget exceeded the cumulative error in 1988 and 1997 was greater than the amount by which the residual of the sodium budget exceeded the cumulative error: -8.15 versus -1.305 kiloequivalents per year in 1988 and -14.05 versus -3.045 kiloequivalents per year in 1997. Further investigation is indicated by the budget for chloride in regard to why the residual of the budget exceeds the cumulative error in just over half the years of the study.

In the case of ANC and DIC, those years in which the residual exceeds the cumulative error suggest there might be an unaccounted for source of alkalinity generation within the lake. Yet, as in the case of the chloride budget, the absence of a consistent excess in each year indicates more information is needed to understand the processes or source of alkalinity that could account for the excess.

The relation of the residual of the solute budget and the cumulative error in the solute budget provides us with an interesting question. In the cases where cumulative error is greater than the residual (calcium, magnesium, sodium, potassium, and DOC), would it be worth the effort in time and cost to focus effort on reducing the cumulative error? By doing so, might additional understanding about patterns or processes in the processing of solutes in the ecosystem be revealed? The same question applies to the cases where no consistent relation exists between the cumulative error and the residual (sulfate, chloride, ANC, and DIC). One thing these solutes have in common is that groundwater fluxes contribute the most to the cumulative error in the solute budget for these solutes (Table 5-11), accounting for over 85 percent of the error (examine Table 5-11 for sulfate, as an example). Based on the average for the 20 years, groundwater input represented 11.9 percent, and groundwater output represented 73.4 percent, for a total of 85.3 percent of the cumulative error in the solute budget for sulfate. The emphasis in this 20-year study has been to determine the groundwater contribution to water fluxes of the lake using a variety of methods: a network of observation wells, groundwater flow models, and the balances for magnesium and stable isotopes (chapter 4). Owing to the fact that these multiple methods provide estimates of groundwater fluxes that are similar in value (chapter 4), it is difficult at this point to identify some further refinement in groundwater fluxes that would reduce those groundwater flux contributions to the cumulative error in both water (Table 5-1) and solute fluxes (Table 5-11).

Let us assume that by using multiple approaches to quantify groundwater fluxes that our estimate of uncertainty for groundwater fluxes, $\pm$ 25 percent, is either underestimated or overestimated. What if this uncertainty were more in the range of $\pm$ 50 percent, or $\pm$10 percent, or even $\pm$ 5 percent? For the purpose of this hypothetical situation, how might this affect our ability to distinguish the solute budget residual from cumulative error in the solute budget? We believe that magnesium is the solute most likely to behave conservatively in Mirror Lake. Magnesium is also an example of a solute for which the solute budget residual is commonly indistinguishable from cumulative error in the solute budget (Table-5-11). An assumed uncertainty in groundwater fluxes of 50 percent produces a similar result to that used in our calculations where uncertainty was assigned a value of 25 percent (Table 5-12); in all years, the absolute value of the residual of the dissolved magnesium budget is less than the cumulative error in the solute budget. With an assumed uncertainty value in groundwater fluxes of 10 percent, or even 5 percent, nearly half of the years are ones in which the absolute values of the solute budget residuals for magnesium are less than the cumulative error.

Change in storage, solute mass in the lake, for phosphorus, nitrate, and ammonium contributed substantially to the cumulative solute budget error. Phosphorus, nitrate, and ammonium were the solutes for which the residual was always or nearly always greater than the cumulative error. This result is of interest, as it may reflect the fact that these solutes exhibit a definite seasonal trend in the lake (chapter 3) and that sample collection in the lake was less frequent than for precipitation and the tributaries. Also, unlike all the other solutes except hydrogen, the precipitation contribution to cumulative error in the solute budget was more than 0.1 percent, being most substantial in the case of nitrate, where the annual average contribution was greater than 50 percent.

BUDGET PATTERNS AND PROCESSES

What have we gained by our examination of uncertainty? Patterns and processes revealed by the water and solute budgets of Mirror Lake have been documented elsewhere, particularly in Likens (1985). Yet that summation recognized that refinements were highly desirable, particularly for groundwater fluxes. In the absence of uncertainty analysis, it is possible that the residual of the solute budget, commonly expected to represent net ecosystem flux, may be either underestimated or overestimated. With uncertainty analysis, comparison of the residual of the solute budget with cumulative error enables us to determine when the residual of the solute budget exceeds cumulative error, such that the excess may better represent an ecosystem process affecting solute mass in the lake.

It is useful, therefore, to reiterate some of the main points of our budget analysis. First, the balance between solute input and loss by hydrologic fluxes for Mirror Lake indicated that the lake was expected to store phosphate, nitrate, ammonium, dissolved silica, and hydrogen ion (chapter 3, Figs. 3-8 and 3-14, Table 5-10). Solute inputs from hydrologic fluxes were substantially larger than solute losses from hydrologic fluxes for these solutes. Concentrations within the lake of phosphate, nitrate, ammonium, and hydrogen ion declined with time (chapter 3). The independently measured change in solute mass in the lake did not correspond to the expected change in mass on the basis of inputs and outputs (chapter 3). Change in mass is affected by concentration changes over time as well as changes in the water mass of the lake. Placing change in solute mass storage in the lake in relation to the hydrologic-flux solute balance results in the residual of the solute budget. The residual of the solute budget provides some indication of gains or losses of solutes within the lake that may result from processes occurring within the lake or some unforeseen or measured hydrologic flux supplying or removing solutes from the lake. Yet uncertainty in the measured hydrological fluxes supplying or removing solutes from the lake, as well as uncertainty in the measurement of change in water volume of the lake, may account for part, or all, of the residual of the solute budget of the lake (Table 5-10). Consequently, for those solutes with values of the residual of the solute budget larger than the cumulative error in the solute budget, the solute mass that may reflect other processes—particulate input, gaseous flux, biotic uptake, or movement—is the value remaining when the cumulative error is subtracted from the residual (Table 5-13). Thus, over the 20 years, apart from an unknown hydrologic flux and ignoring the effect of sample collection frequency in the lake on determination of change in lake solute mass, processes within the lake must remove something on the order of 2.0 kg yr^{-1} of phosphate, 257 kg yr^{-1} of nitrate, 36.1 kg yr^{-1} of ammonium, 2395 kg yr^{-1} dissolved silica, and 9050 eq yr^{-1} of hydrogen ion.

THE RELATION OF WATER AND SOLUTE BUDGETS TO LAKE CONCENTRATIONS

One other way to examine how well water and solute inputs and outputs have been determined is to use those data to estimate the average solute concentration in the lake. The estimated concentrations can be compared to the independently determined average lake concentrations. Where there is good agreement, it is likely that all of the inputs and outputs have been accounted for, that they are well constrained in value, and that the model is representative of the controls on lake concentration.

For the purposes of this analysis, we chose to use the nutrient budget model developed by Dillon and Rigler (1974). In this model, the average concentration of solute in the lake ($[Cx]$) is determined from solute loading (Lx), solute retention (Rx), lake mean depth (z), and hydraulic flushing rate (ρ). Hydraulic flushing rate (ρ) is calculated by dividing total water outflow, surface water, and ground water by lake volume. Note retention as used here simply refers to the proportion of solute mass input not lost by solute mass losses associated with hydrologic influx and outflux. Thus, the calculation estimates what the concentration of solute should be in the lake on the basis of measurement of the amount of solute entering the lake and exiting the lake as well as how fast water is moving through the lake.

$$[Cx] = Lx\,(1-Rx)/z\rho$$

The results of this model are in good agreement with the average concentrations in the lake, based on examination of the statistical properties of 20 years of measured concentrations and 20 years of model estimates, except in the case of hydrogen ion, where only 10 years of volume-weighted average values were available, and nitrate concentration, which was overestimated some 6.8-fold by the model. Hydrogen ion (H^+) was overestimated some 1.6-fold by the model (Table 5-14). The discrepancy may reflect the fact that the model is based on a hydrological characteristic, hydraulic flushing rate, and on the difference between chemical mass input and output associated with hydrological fluxes. Processes within the lake that remove nitrate and hydrogen from the water column of the lake are not included in the model. The net difference among the hydrologic fluxes may indicate that a certain change in mass in the lake should result from that difference, yet biological uptake within the lake can remove the solute from solution, thereby decreasing the concentration of the solute in the lake. The relation of the estimated values to the actual measured volume-weighted averages is also of interest when individual years are compared over the course of the study. The next series of figures shows these relations. The trend in concentration was also well-represented for selected solutes (Fig. 5-6). The good agreement between concentrations estimated from water and solute inputs and outputs and the measured volume-weighted average concentrations in the lake, with the exception of nitrate and hydrogen ion, reinforce the

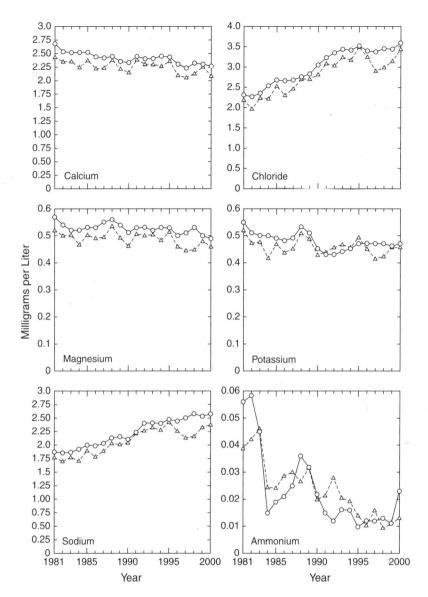

FIGURE 5-6.1. Comparison of measured annual average concentrations in Mirror Lake with annual average concentrations estimated from a simple chemical model that uses solute mass loading, solute retention, and hydraulic flushing rate..

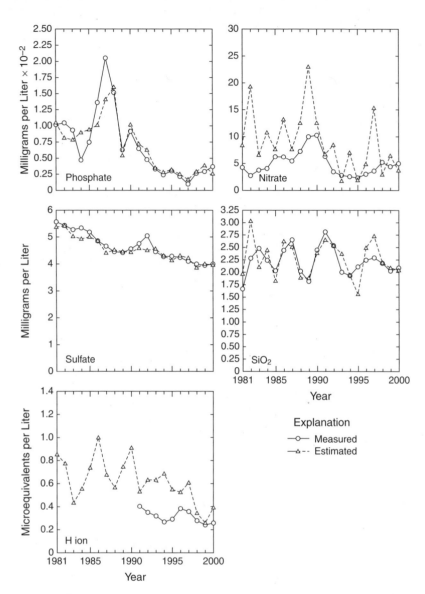

FIGURE 5-6.2.

fact that the water and solute budgets of Mirror Lake are generally well constrained. The reader is cautioned, however, that in the case of Mirror Lake, sufficient multiple lines of evidence exist that the water budgets are well defined. Thus, the simple solute model results presented herein add further evidence in support of our conclusion, rather than representing a test of those conclusions apart from other evidence.

The existing literature in which groundwater and/or solute fluxes have been quantified in lakes indicates that the groundwater contribution can vary from negligible (Malueg et al. 1975; Schindler et al. 1976; Sacks 2002) to substantial (Brown and Cherkauer 1992; LaBaugh et al. 1995; Sacks 2002). Also, even in a lake that receives modest influx of water from ground water (less than 10 percent of total water inputs), solute influx from ground water can be an important regulator of in-lake processes (Wentz et al. 1995). In the case of the Wentz et al.'s (1985) eight-year study, changes in precipitation from relatively wet to dry conditions resulted in groundwater influx declining from 7 percent of total influx to the point where it ceased. The change in precipitation resulted in a decline in the solute contribution to the lake from ground water, from nearly half or more of total solute inputs in the wet years to not at all in the dry years. Similarly, Krabbenhoft and Webster (1995), in a four-year study, found dry conditions resulted in a cessation of ground water solute and water influx resulting in increased acidification of a lake. In the case of Mirror Lake, considerable attention was given to assessing the groundwater contribution to the water and solute fluxes, balances, and budgets of the lake. One outcome of this book may be to encourage others who need to quantify water and solute fluxes for a lake or lakes to at least consider how best to account for all water fluxes, including ground water, and what the result might be if they do not.

One common approach to the estimation of groundwater flux contributions to lakes has been that of determining net groundwater flux from the difference between measured water influx and water loss from a lake (Winter 1981; Sacks et al. 1998). As noted in those studies, the net flux does not enable the determination of absolute groundwater fluxes into or out of a lake, and the net value obtained from the difference of measured water fluxes is affected by uncertainty of measurement of those fluxes.

If the annual groundwater contribution to Mirror Lake had been assumed to be the net value obtained by subtracting the annual volumes of surface outlet, evaporation, and change in storage from precipitation and tributary fluxes, then the net groundwater flux would have been thought to be positive, with more ground water coming into the lake than going out of the lake. The actual difference between groundwater input and groundwater output, however, was negative. In each of the 20 years of study, groundwater output exceeded groundwater input (Table 5-15).

UNCERTAINTY IN PERSPECTIVE

> In physical science the first essential step in the direction of learning any subject is to find principles of numerical reckoning and practicable methods for measuring some quality connected with it. I often say that when you can measure what you are speaking about, and express it in numbers, you know something about it; but when you cannot measure it, when you cannot express it in numbers, your knowledge is of a meagre and unsatisfactory kind; it may be the beginning of knowledge, but you have scarcely in your thoughts advanced to the state of Science, whatever the matter may be.
>
> LORD KELVIN 1883

In the case of Mirror Lake, our efforts to measure the water and chemical budgets, including emphasis on quantification of groundwater fluxes, provide not only a beginning point for the discussion of pattern and process in the lake but also some advance in the state of our knowledge about the lake and its watershed and airshed. In the previous Mirror Lake book (Likens 1985), the need to focus further attention on groundwater fluxes was acknowledged. In this book, we find that use of multiple lines of evidence to better define groundwater fluxes has produced a coherent water budget for the lake. Uncertainty exists in the budget, but as noted in chapter 4 and this chapter, the absolute value of that uncertainty gives us confidence that we have not overlooked processes controlling the basic hydrological characteristics of the lake. Furthermore, such uncertainty analysis provides an indication that for some solutes, other processes may affect solute concentrations in the lake, and the amount by which the residual exceeds cumulative error in a budget is likely to be an indication of the amount of solute attributable to those processes.

Having some measure of uncertainty also provides a beginning in relation to how much more might need to be done to understand controls on pattern and processes affecting solutes in any ecosystem. In the absence of uncertainty analysis, the investigator may interpret the residual of the budget as representing solutes unaccounted for by hydrological processes alone. Yet we know that some of the residual may be a result of uncertainty in measurements of the chemical budget. Use of uncertainty analysis, therefore, provides an expression in numbers of what part of the residual may be attributable to measurement uncertainty, rather than being a function of some environmental process affecting solutes apart from hydrological fluxes. No matter what the relation between the budget residual and cumulative error, the investigator may need to ask the following question: are the budgets well-constrained, measured with the best available methods (those associated with the least uncertainty), or might some improvement in measurement of one or all of the hydrological fluxes provide more useful information about controls on solutes in the lake or ecosystem? This chapter indicates that being able to express uncertainty in water and solute budgets aids in advancing knowledge about the meaning of and confidence in water and solute budgets in relation to environmental processes and their management.

TABLE 5-1. *Volume represented by the residual of the water balance and cumulative error in measured water balance terms for Mirror Lake, 1981 to 2000*

Year	Water-balance residual, in $m^3 \times 1000$	Cumulative error, in $m^3 \times 1000$
1981	86.2	111.5
1982	−10.1	99.6
1983	124.6	92.6
1984	124.2	97.4
1985	48.0	84.6
1986	107.8	98.5
1987	−16.6	105.0
1988	−88.8	105.0
1989	7.6	103.3
1990	235.4	98.5
1991	5.5	102.4
1992	−2.2	98.5
1993	−96.5	103.7
1994	−52.9	99.9
1995	30.2	89.3
1996	94.0	96.6
1997	−85.6	109.0
1998	45.4	97.3
1999	−17.2	103.8
2000	21.8	98.5

TABLE 5-2. *First-order error values, variance in cumulative error, and percent contribution to cumulative error for Mirror Lake water budget components, 1981 to 2000*

The symbol δ^2 represents the variance in the cumulative error in the water budget, and δ_i^2 represents the error in individual water budget measurements as shown in Equation 2–3 in Chapter 2.

Year	Precipitation δ^2_P	Evaporation δ^2_E	Tributaries δ^2_{SWI}
1981			
First-order error value (1000 m³)	126.788	130.645	839.55
Annual contribution to cumulative error	1.0%	1.1%	6.8%
1982			
First-order error value (1000 m³)	57.836	130.645	329.42
Annual contribution to cumulative error	0.6%	1.3%	3.3%
1983			
First-order error value (1000 m³)	98.208	93.896	440.79
Annual contribution to cumulative error	1.1%	1.1%	5.1%
1984			
First-order error value (1000 m³)	87.797	96.531	587.34
Annual contribution to cumulative error	0.9%	1.0%	6.2%
1985			
First-order error value (1000 m³)	60.918	150.185	228.77
Annual contribution to cumulative error	0.9%	2.1%	3.2%
1986			
First-order error value (1000 m³)	88.642	139.004	543.36
Annual contribution to cumulative error	0.9%	1.4%	5.6%

Surface outlet δ^2_{SWO}	Ground water input δ^2_{GWI}	Ground water output δ^2_{GWO}	Change in lake mass $\delta^2_{\Delta L}$	Cumulative error variance δ^2
4515.84	833.766	5986.89	.0016	12,433.48
36.3%	6.7%	48.2%	<1%	
1074.20	832.323	7490.90	.0009	9915.33
10.8%	8.4%	75.5%	<1%	
1292.76	835.210	5806.44	.0001	8567.31
15.1%	9.7%	67.8%	<1%	
2006.14	915.063	5795.02	.0004	9487.89
21.1%	9.6%	61.1%	<1%	
496.84	803.723	5409.60	.0025	7150.04
6.9%	11.2%	75.7%	<1%	
1513.99	836.656	6573.16	.0001	9694.80
15.6%	8.6%	67.8%	<1%	

(continued)

TABLE 5-2 (CONTINUED).

Year	Precipitation δ^2_P	Evaporation δ^2_E	Tributaries δ^2_{SWI}
1987			
First-order error value (1000 m³)	66.994	157.252	280.06
Annual contribution to cumulative error	0.6%	1.4%	2.5%
1988			
First-order error value (1000 m³)	46.854	118.592	140.54
Annual contribution to cumulative error	0.4%	1.1%	1.3%
1989			
First-order error value (1000 m³)	92.833	173.449	425.80
Annual contribution to cumulative error	0.9%	1.6%	4.0%
1990			
First-order error value (1000 m³)	133.749	123.877	1002.99
Annual contribution to cumulative error	1.4%	1.3%	10.3%
1991			
First-order error value (1000 m³)	82.537	133.056	400.20
Annual contribution to cumulative error	0.8%	1.3%	3.8%
1992			
First-order error value (1000 m³)	73.616	114.062	401.80
Annual contribution to cumulative error	0.8%	1.2%	4.1%
1993			
First-order error value (1000 m³)	61.466	124.881	267.49
Annual contribution to cumulative error	0.6%	1.2%	2.5%

Surface outlet δ^2_{SWO}	Ground water input δ^2_{GWI}	Ground water output δ^2_{GWO}	Change in lake mass $\delta^2_{\Delta L}$	Cumulative error variance δ^2
628.25	841.000	9044.01	.0001	11,017.57
5.7%	7.6%	82.1%	<1%	
214.33	677.301	9835.68	.0001	11,033.30
1.9%	6.1%	89.1%	<1%	
1550.39	689.063	7744.00	.0009	10,675.54
14.5%	6.5%	72.5%	<1%	
1413.20	894.010	6126.98	.0004	9694.80
14.6%	9.2%	63.2%	<1%	
142.21	843.903	8887.78	.0001	10,489.68
1.4%	8.0%	84.7%	<1%	
158.13	757.626	8199.30	.0000	9704.54
1.6%	7.8%	84.5%	<1%	
126.56	724.956	9447.84	.0009	10,753.19
1.2%	6.7%	87.9%	<1%	

(continued)

TABLE 5-2 (CONTINUED).

Year	Precipitation δ^2_P	Evaporation δ^2_E	Tributaries δ^2_{SWI}
1994			
First-order error value (1000 m³)	65.448	151.290	320.59
Annual contribution to cumulative error	0.7%	1.5%	3.2%
1995			
First-order error value (1000 m³)	80.730	132.020	314.35
Annual contribution to cumulative error	1.0%	1.7%	3.9%
1996			
First-order error value (1000 m³)	140.304	295.496	1067.33
Annual contribution to cumulative error	1.5%	3.2%	11.4%
1997			
First-order error value (1000 m³)	68.724	95.063	362.14
Annual contribution to cumulative error	0.6%	0.8%	3.0%
1998			
First-order error value (1000 m³)	87.703	123.210	471.76
Annual contribution to cumulative error	0.9%	1.3%	5.0%
1999			
First-order error value (1000 m³)	73.188	128.596	354.76
Annual contribution to cumulative error	0.7%	1.2%	3.3%
2000			
First-order error value (1000 m³)	78.677	84.548	462.25
Annual contribution to cumulative error	0.8%	0.9%	4.8%

Surface outlet δ^2_{SWO}	Ground water input δ^2_{GWI}	Ground water output δ^2_{GWO}	Change in lake mass $\delta^2_{\Delta L}$	Cumulative error variance δ^2
133.29	756.250	8547.00	.0000	9973.87
1.3%	7.6%	85.7%	<1%	
99.80	623.751	6715.80	.0001	7966.46
1.3%	7.8%	84.3%	<1%	
586.37	913.551	6320.25	.0016	9323.30
6.3%	9.8%	67.8%	<1%	
196.42	876.160	10276.89	.0016	11,875.40
1.7%	7.4%	86.5%	<1%	
164.35	743.926	7872.13	.0001	9463.08
1.7%	7.9%	83.2%	<1%	
107.43	700.926	9404.15	.0000	10,769.05
1.0%	6.5%	87.3%	<1%	
189.48	775.623	8113.51	.0000	9704.08
2.0%	8.0%	83.6%	<1%	

TABLE 5-3. *Results of least-squares regression analysis relating chemical concentration and the daily flow of the day on which sample collection occurred for those chemicals for which the relation was statistically significant*

Water budget component	Observations	Intercept	Slope	R^2	F value	$P > F$[a]
Calcium (1981–2000)						
Surface outlet	1088	2.52	−0.0933	0.051	58.2	<0.001
Northeast inlet	1025	15.28	−4.092	0.420	740.5	<0.001
Northwest inlet	1096	5.35	−1.134	0.602	1655.5	<0.001
West tributary	1093	5.47	−1.154	0.588	1559.8	<0.001
Magnesium (1981–2000)						
Surface outlet	1088	0.56	−0.0266	0.080	94.4	<0.001
Northeast stream	1024	3.38	−0.935	0.454	848.6	<0.001
Northwest stream	1096	1.26	−0.286	0.670	2216.9	<0.001
West tributary	1093	1.21	−0.260	0.603	1660.2	<0.001
Dissolved silica (1981–2000)						
Surface outlet	1088	1.15	0.420	0.166	215.4	<0.001
Northeast stream	1025	19.63	−5.226	0.739	2902.0	<0.001
Northwest stream	1096	20.47	−4.742	0.766	3578.7	<0.001
West tributary	1093	21.35	−4.936	0.767	3589.0	<0.001
Sodium (1981–1990)						
Surface outlet	529	2.17	−0.112	0.064	35.8	<0.001
Northeast tributary	517	26.11	−3.692	0.070	39.0	<0.001
Northwest tributary	528	3.94	−0.913	0.716	1328.4	<0.001
West tributary	528	3.98	−0.758	0.539	615.5	<0.001
Sodium (1991–2000)						
Surface outlet	559	2.73	−0.164	0.181	122.9	<0.001
Northeast tributary	508	44.67	−9.458	0.237	157.3	<0.001
Northwest tributary	568	5.33	−1.133	0.703	1342.9	<0.001
West tributary	565	5.59	−1.211	0.567	736.0	<0.001

TABLE 5-3 (CONTINUED).

Water budget component	Observations	Intercept	Slope	R^2	F value	$P > F$[a]
		Chloride (1981–1990)				
Surface outlet	529	2.98	−0.208	0.77	44.0	<0.001
Northeast tributary	516	66.59	−13.002	0.175	108.9	<0.001
Northwest tributary	528	2.27	−0.459	0.271	195.2	<0.001
West tributary	528	3.59	−0.572	0.133	80.5	<0.001
		Chloride (1991–2000)				
Surface outlet	559	3.89	−0.259	0.175	118.1	<0.001
Northeast tributary	508	99.3	−24.865	0.333	252.4	<0.001
Northwest tributary	568	4.95	−1.246	0.464	490.7	<0.001
West tributary	565	6.06	−1.20	0.232	170.1	<0.001

NOTE: The regression model was solute concentration = intercept ± slope * (Log10 flow).

[a] Probability of a greater F value due to chance; relation between concentration and flow is statistically significant at the 0.01 level.

TABLE 5-4. *Summary of the comparison of chemical mass over 20 years*

Water budget component	Mass-flux calculation	Number of observations	Average	Standard deviation	Minimum	Maximum
		Calcium				
Surface outlet	Model	20	566.2	195.8	220.3	1055.8
	Average	20	564.9	200.9	219.3	1078.3
	% difference	20	0.5	4.9	−6.3	12.2
Northeast tributary	Model	20	97.2	22.3	63.1	144.0
	Average	20	100.0	22.7	64.0	141.3
	% difference	20	−2.2	11.0	18.2	25.1
Northwest tributary	Model	20	478.0	95.5	324.6	696.0
	Average	20	498.2	118.0	345.6	752.4
	% difference	20	−3.2	6.9	−14.7	9.8
West tributary	Model	20	346.7	68.1	223.2	485.8
	Average	20	355.2	70.3	244.2	520.9
	% difference	20	−2.3	4.5	−10.0	7.2
		Magnesium				
Surface outlet	Model	20	121.4	41.8	47.6	226.2
	Average	20	121.3	43.3	48.6	231.7
	% difference	20	0.4	4.8	−6.6	11.8
Northeast tributary	Model	20	20.5	4.6	13.4	30.1
	Average	20	21.4	4.3	15.7	29.5
	% difference	20	−3.7	10.7	−18.3	21.0
Northwest tributary	Model	20	98.8	18.3	69.4	141.1
	Average	20	103.4	22.9	75.1	153.1
	% difference	20	−3.7	6.1	−14.1	6.6
West tributary	Model	20	75.2	14.5	48.7	104.9
	Average	20	77.1	14.5	54.1	111.8
	% difference	20	−2.6	4.0	−10.0	4.4
		Sodium				
Surface outlet	Model	20	513.1	185.3	181.7	1053.6
	Average	20	514.8	197.1	185.4	1094.4
	% difference	20	0.4	6.2	−17.0	12.1
Northeast tributary	Model	20	286.1	89.0	144.0	515.3
	Average	20	282.1	95.4	132.3	495.1
	% difference	20	3.6	16.3	−23.1	44.7
Northwest tributary	Model	20	309.4	55.2	205.6	422.2
	Average	20	329.6	83.3	209.0	528.6
	% difference	20	−4.5	7.9	−20.9	7.3
West tributary	Model	20	314.4	68.2	177.9	477.6
	Average	20	326.8	88.2	182.8	519.3
	% difference	20	−2.4	7.9	−20.0	11.3

TABLE 5-4 (CONTINUED).

Water budget component	Mass-flux calculation	Number of observations	Average	Standard deviation	Minimum	Maximum
		Chloride				
Surface outlet	Model	20	687.2	259.4	234.8	1468.8
	Average	20	694.1	286.1	243.7	1570.1
	% difference	20	0.4	9.7	−22.1	27.3
Northeast tributary	Model	20	596.2	162.9	329.4	1004.3
	Average	20	596.2	179.3	311.2	1004.6
	% difference	20	1.7	14.4	−20.4	37.3
Northwest tributary	Model	20	257.7	59.4	145.2	379.4
	Average	20	276.7	102.3	141.5	517.5
	% difference	20	−2.4	15.3	−32.5	25.3
West tributary	Model	20	363.5	95.1	185.6	598.9
	Average	20	375.8	142.4	188.3	692.4
	% difference	20	0.8	14.9	−32.3	33.6
		Dissolved silica				
Surface outlet	Model	20	638.7	229.6	227.6	1241.8
	Average	20	621.2	234.6	207.8	1215.4
	% difference	20	5.2	15.4	−25.2	45.8
Northeast tributary	Model	20	125.8	30.2	78.5	188.8
	Average	20	134.1	35.5	79.0	206.2
	% difference	20	−5.7	3.5	−12.8	1.3
Northwest tributary	Model	20	1523.2	280.3	1061.1	2160.9
	Average	20	1605.6	381.9	1019.4	2400.0
	% difference	20	−4.0	6.2	−16.4	4.9
West tributary	Model	20	1161.3	218.9	759.1	1624.2
	Average	20	1215.7	269.4	746.6	1894.7
	% difference	20	−3.8	4.5	−14.3	2.9

NOTE: Average concentrations were determined from analyses obtained from weekly sample collection (average), and chemical mass was determined using concentrations determined from regression analysis of concentration and flow (model). Data are presented for those chemicals for which a statistically significant relation existed between concentration and flow, and those relations were used to calculate alternate fluxes.

TABLE 5-5. *Budget residuals of inputs, outputs, and change in storage, in kilograms/year*

Year	Calcium		Magnesium	
	Average	Model	Average	Model
1981	326.9	241.1	36.3	22.9
1982	−36.2	−76.8	6.5	−6.2
1983	263.0	218.4	49.6	41.3
1984	200.4	63.5	30.6	5.6
1985	289.0	261.0	39.7	37.8
1986	252.0	197.7	9.9	2.8
1987	55.8	−1.0	−7.1	−22.2
1988	−131.0	−185.2	−50.0	−62.4
1989	282.9	150.8	53.2	23.8
1990	199.1	178.1	52.9	49.5
1991	−16.4	−10.9	−13.7	−17.3
1992	6.2	5.0	−3.7	−10.8
1993	−262.6	−256.8	−50.3	−50.2
1994	−110.6	−157.8	−31.6	−42.8
1995	74.7	67.8	−1.3	−1.0
1996	209.9	187.2	38.4	29.4
1997	−133.9	−130.6	−83.2	−88.0
1998	−91.1	−94.6	32.1	31.0
1999	15.8	147.8	6.1	11.6
2000	−152.3	−119.0	−26.6	−23.0

NOTE: Data were determined using average concentrations between chemical sample collection dates (average), and concentrations were determined from a regression equation relating concentration to flow (model).

Average: All chemical budget components were calculated according to methods described in chapter 3. Average concentration was used between chemical sample collection dates with daily flow values to calculate chemical mass input from the tributary streams and chemical mass lost through the surface water outlet. Precipitation input, groundwater input and output, and change in storage were calculated as in chapter 3.

Equation: Precipitation, groundwater input and output, and change in storage were calculated as in chapter 3. The rest was calculated using concentration estimated from regression equation.

Sodium		Chloride		Dissolved Silica	
Average	Model	Average	Model	Average	Model
130.2	165.6	−25.6	60.2	3693.1	2802.4
−54.4	−6.0	−383.0	−265.7	1997.3	1860.8
193.2	232.8	−186.1	−117.8	3143.1	3093.3
8.8	23.0	−405.2	−306.5	3519.6	3170.8
158.3	187.6	−176.8	−131.6	2853.4	2823.8
98.8	105.0	−191.8	−125.9	2173.5	2070.1
−47.0	−64.3	−408.9	−400.7	2895.5	2766.5
−251.8	−287.9	−568.1	−631.0	2744.3	2778.1
134.3	−8.5	−212.7	−424.2	2552.9	2284.6
435.2	328.6	−112.2	−248.9	3102.4	2838.1
−210.1	−191.9	−266.8	−235.7	2777.0	2775.7
−156.6	−153.9	−632.2	−636.8	2768.6	2805.7
−148.9	−133.8	−467.7	−493.7	2577.5	2650.0
−227.0	−267.4	−644.2	−693.2	2092.6	2031.5
−72.6	−98.7	−424.6	−496.5	3093.4	2838.3
198.8	107.8	50.9	−28.7	3185.7	2772.7
−321.4	−349.7	−822.9	−813.4	2379.9	2531.4
107.0	57.5	−175.9	−184.2	2917.0	2644.5
59.3	−29.1	−360.4	−503.1	3088.0	3024.2
−125.6	−233.5	−435.6	−658.6	2486.4	2226.5

TABLE 5-6. *Two methods of determining chemical budgets of surface-water contributions*

Method used to calculate surface water inlet and outlet mass flux	Shapiro-Wilk statistic W-value	Probability of a smaller value of W	Distribution significantly different than a normal distribution	t value for test that variances are equal	Probability of a larger absolute value of t	Is residual value variance different?
Calcium							
Average	0.93	0.21	No	0.61	0.54	No
Model	0.94	0.26	No			
Magnesium							
Average	0.94	0.24	No	0.63	0.53	No
Model	0.94	0.25	No			
Sodium							
Average	0.96	0.71	No	0.45	0.65	No
Model	0.98	0.97	No			

	Chloride			
Average	0.97	0.89	0.33	0.74	No
Model	0.96	0.72			
			X^2 value for test that data have a common median	Probability of a larger value of X^2	Do distributions share a median?

	Dissolved silica			
Average	0.97	0.90	0.39	0.53	Yes
Model	0.89	0.03			

NOTE: Data were determined in two different ways. The chemical budget characteristic subject to analysis was the residual of all measured terms: inputs minus outputs minus change in storage. The number of observations for all tests was 20.

TABLE 5-7. *Comparison of chemical retention on an annual basis in Mirror Lake using two different methods to determine percentage of chemical-mass input from surface water to Mirror Lake and percentage of chemical-mass loss through the surface outlet*

	N	Mean	Median	Standard deviation	Minimum	Maximum
			Calcium			
Average	20	2.235	1.20	10.166	−17.5	22.0
Model	20	0.105	−1.10	9.632	−17.1	21.6
			Magnesium			
Average	20	−0.775	−1.70	9.979	−20.2	18.9
Model	20	−3.345	−5.15	10.218	−21.1	18.6
			Sodium			
Average	20	1.405	0.95	10.816	−21.4	23.7
Model	20	−0.445	−0.90	11.868	−20.2	20.4
			Chloride			
Average	20	−22.86	−20.80	15.462	−51.2	11.6
Model	20	−24.155	−21.60	17.414	−63.5	6.3
			Dissolved silica			
Average	20	67.22	68.80	6.481	52.6	78.0
Model	20	65.835	66.60	4.917	55.7	75.8

NOTE: The first method (average) was as described and presented in chapter 3—the average concentrations between sample collection dates was used to calculate chemical mass input from surface water and chemical mass loss by surface water. The second method (model) used concentration determined from the regression model relating concentration to flow to determine surface water chemical inputs and outputs.

Chemical retention = (1 − output/input) × 100.

TABLE 5-8. *Concentrations used to represent groundwater chemical characteristics in the calculation of chemical mass input to Mirror Lake from ground water*

Value used in calculation	Ca (mg/l)	Mg (mg/l)	K (mg/l)	Na (mg/l)	NH_4 (mg/l)	H^+ (µeq/l)	SO_4 (mg/l)	NO_3 (mg/l)	Cl (mg/l)	PO_4 (mg/l)	DOC (mg/l)	DIC (µM/l)	ANC (µeq/l)	Dissolved silica (mg/l)
Median of all wells	3.93	0.79	1.24	2.90	0.020	0.22	5.63	0.008	1.10	0.0030	0.88	715	320.0	11.20
Median of low stream flow	3.34	0.68	0.65	2.68	0.010	0.20	4.59	0.008	2.38	0.0080	2.76	210	165.3	11.93
Median of flow at constant concentration	3.59	0.79	0.66	2.58	0.020	0.17	4.50	0.065	2.18	0.0073	2.62	213	185.2	12.30

TABLE 5-9. *Comparison of chemical retention on an annual basis in Mirror Lake using three different methods to determine percentage of chemical mass input from ground water to Mirror Lake*

Method	N	Mean	Median	Standard deviation	Minimum	Maximum
Calcium						
Median	20	2.235	1.20	10.166	−17.5	22.0
Low flow	20	−2.710	−3.15	11.012	−24.0	19.0
Constant	20	−0.550	−1.25	10.637	−21.2	20.3
Magnesium						
Median	20	−0.775	−1.70	9.979	−20.2	18.9
Low flow	20	−5.280	−6.00	10.672	−26.1	16.0
Constant	20	−0.775	−1.70	9.979	−20.2	18.9
Sodium						
Median	20	1.405	0.95	10.816	−21.4	23.7
Low flow	20	−0.570	−0.75	11.100	−23.9	22.6
Constant	20	−1.515	−1.60	11.229	−25.1	22.1
Potassium						
Median	20	9.740	8.80	8.247	−6.0	23.2
Low flow	20	−15.120	−15.10	10.477	−35.8	5.6
Constant	20	−14.585	−14.60	10.403	−35.1	5.9
Chloride						
Median	20	−22.860	−20.80	15.462	−51.2	11.6
Low flow	20	−10.550	−9.10	12.600	−31.8	17.5
Constant	20	−12.305	−10.85	12.945	−34.1	16.6
Sulfate						
Median	20	14.075	14.95	9.540	−4.0	34.3
Low flow	20	10.725	11.65	10.242	−8.5	32.4
Constant	20	10.425	11.35	10.326	−9.0	32.3
Acid-neutralizing capacity						
Median	9	20.767	21.8	7.353	9.1	28.2
Low flow	9	−13.711	−10.8	11.849	−33.8	−0.7
Constant	9	−7.667	−5.1	10.995	−26.1	4.2

TABLE 5-9 (CONTINUED).

Method	N	Mean	Median	Standard deviation	Minimum	Maximum
			Hydrogen ion			
Median	20	96.075	95.8	1.286	92.2	97.7
Low flow	20	96.075	95.8	1.286	92.2	97.7
Constant	20	96.075	95.8	1.286	92.2	97.7
			Dissolved silica			
Median	20	67.22	68.8	6.481	52.6	78.0
Low flow	20	67.875	69.45	6.333	53.6	78.5
Constant	20	68.18	69.75	6.265	54.1	78.7
			Phosphate			
Median	20	39.725	35.15	21.910	1.2	82.5
Low flow	20	44.945	38.85	20.372	9.2	83.3
Constant	20	44.245	38.30	20.590	8.1	83.2
			Nitrate			
Median	20	83.835	85.05	8.940	65.4	96.9
Low flow	20	84.225	85.45	8.767	66.0	97.0
Constant	20	84.145	85.35	8.794	65.9	97.0
			Ammonium			
Median	20	71.855	72.20	13.726	42.3	88.0
Low flow	20	71.195	71.55	14.085	40.8	87.8
Constant	20	71.86	72.35	13.723	42.3	88.0
			Dissolved organic carbon			
Median	4	2.525	1.25	12.092	−10.6	18.2
Low flow	4	16.625	16.15	8.984	6.2	28.0
Constant	4	15.725	15.20	9.206	5.1	27.4
			Dissolved inorganic carbon			
Median	7	36.728	37.1	5.742	27.5	46.4
Low flow	7	−8.414	−9.3	11.862	−25.8	10.1
Constant	7	−7.957	−8.8	11.806	−25.3	10.5

NOTE: The first method (median) was as described and presented in chapter 3; the median concentration found in ground water was used to calculate chemical mass input from ground water. The second method (low flow) used the median concentration at low flow. The third (constant) used median concentration at low flow when concentration was constant.

Chemical retention = $(1 - \text{output/input}) \times 100$.

TABLE 5-10. *Comparison of the residual and the cumulative error of the solute budgets for Mirror Lake, 1981 to 2000*

Bold type indicates cumulative error value is larger than absolute value of the residual for that year.

Year	Calcium (kg/yr) Residual	Cumulative error	Magnesium (kg/yr) Residual	Cumulative error
1981	326.9	285.1	36.2	**60.1**
1982	−36.3	**253.5**	6.5	**53.4**
1983	263.0	234.2	49.6	49.4
1984	200.4	**239.2**	30.6	**49.5**
1985	289.0	218.1	39.7	**45.5**
1986	252.0	237.6	10.0	**51.4**
1987	55.9	**253.7**	−7.2	**55.2**
1988	−131.0	**263.3**	−50.0	**58.8**
1989	282.9	251.8	53.3	**55.0**
1990	199.2	**232.7**	52.9	49.0
1991	−16.3	**258.7**	−13.7	**54.3**
1992	6.3	**243.1**	−3.6	**52.1**
1993	−262.6	254.9	−50.3	**55.2**
1994	−110.7	**247.7**	−31.7	**52.7**
1995	74.7	**221.9**	−1.3	**47.7**
1996	209.9	**220.8**	38.3	**47.2**
1997	−133.8	**250.6**	−83.2	53.9
1998	−91.1	**225.5**	32.2	**47.1**
1999	115.7	**248.9**	6.0	**52.5**
2000	−152.3	226.9	−26.6	**49.0**

	Sodium (kg/yr)		Potassium (kg/yr)		Hydrogen ion (eq/yr)	
Residual	Cumulative error	Residual	Cumulative error	Residual	Cumulative error
130.2	**206.4**	45.5	**65.4**	12,837.7	634.5
−54.4	**185.9**	37.5	**58.6**	9781.4	491.6
193.1	175.3	58.6	54.4	9887.1	483.6
18.8	**182.4**	60.8	53.8	9839.6	489.0
158.3	**170.6**	79.8	50.8	9438.5	477.8
98.9	**188.9**	59.4	54.1	10,279.4	521.3
−47.0	**208.6**	4.0	**58.1**	8166.7	416.5
−251.8	221.8	−35.5	**61.4**	9683.7	494.6
134.3	**220.1**	85.1	59.7	10,487.3	557.5
435.2	210.5	110.8	53.7	12,890.2	643.3
−210.1	**234.4**	49.8	**55.8**	10,478.2	521.5
−156.6	**228.9**	18.7	**54.6**	8149.1	411.3
−148.9	**249.8**	−19.6	**57.8**	8853.5	443.6
−226.9	**241.9**	−1.4	**56.9**	9418.0	472.5
−72.6	**218.0**	1.7	**51.5**	9122.5	459.3
198.8	**221.3**	51.2	**54.1**	9541.4	462.8
−321.3	251.3	2.1	**58.5**	8649.0	429.1
107.0	**218.5**	61.2	52.0	7907.0	385.1
59.3	**247.6**	28.7	**56.1**	6824.1	331.3
−125.6	**243.5**	1.4	**55.5**	8314.9	404.9

(continued)

TABLE 5-10 (CONTINUED).

	Chloride (kg/yr)		Sulfate (kg/yr)	
Year	Residual	Cumulative error	Residual	Cumulative error
1981	−25.5	**235.3**	850.5	602.1
1982	−383.1	197.4	−39.8	**535.3**
1983	−186.1	**198.4**	1195.8	468.7
1984	−405.1	210.4	287.6	**488.7**
1985	−176.8	**202.5**	1338.2	432.1
1986	−191.8	**218.4**	533.6	481.5
1987	−408.9	252.0	887.9	480.6
1988	−568.1	279.0	−192.4	**487.0**
1989	−212.7	**279.1**	968.0	479.8
1990	−112.2	**268.6**	962.3	457.8
1991	−266.9	**304.8**	746.1	481.7
1992	−632.1	289.6	−551.0	461.6
1993	−467.6	333.8	78.9	**482.7**
1994	−644.2	320.5	393.1	**440.8**
1995	−424.6	297.5	509.3	383.2
1996	50.9	**293.2**	974.0	429.6
1997	−822.9	324.5	48.8	**473.1**
1998	−175.8	**284.6**	623.6	393.6
1999	−360.3	320.6	326.9	**423.9**
2000	−435.6	331.7	446.1	409.9

Acid-neutralizing capacity (eq/yr)		Dissolved organic carbon (kg/yr)		Dissolved inorganic carbon (mol/yr)	
Residual	Cumulative error	Residual	Cumulative error	Residual	Cumulative error
.	.	.	.	.	.
.	.	.	.	.	.
.	.	.	.	.	.
.	.	.	.	.	.
.	.	.	.	.	.
.	.	.	.	.	.
.	.	.	.	.	.
.	.	.	.	.	.
.	.	.	.	.	.
19,175.8	10,957.3	.	.	.	.
−1879.9	**11,547.7**	.	.	.	.
10,064.7	**11,622.5**	−17.8	171.9	31,240.7	23,796.2
16,461.3	10,324.1	.	.	41,307.1	21,352.7
35,090.3	10,915.7	.	.	67,305.0	24,366.0
−3031.7	**11,800.8**	.	.	43,906.2	26,074.5
15,562.2	10,994.2	272.8	196.5	71,717.1	22,547.0
3552.0	**11,618.8**	−117.9	**209.2**	58,051.5	22,985.8
9194.6	**11,520.3**	53.7	**197.2**	17,340.3	**23,651.6**

(continued)

TABLE 5-10 (CONTINUED).

Year	Phosphate (kg/yr)		Nitrate (kg/yr)	
	Residual	Cumulative error	Residual	Cumulative error
1981	−1.6	1.1	382.8	19.2
1982	2.6	0.8	218.0	20.5
1983	2.1	0.7	243.8	13.4
1984	8.8	1.1	241.1	16.2
1985	−10.3	1.4	245.2	15.5
1986	3.7	1.0	285.9	18.3
1987	−10.0	1.9	210.0	13.8
1988	23.6	2.7	275.4	18.8
1989	−5.4	0.9	223.3	27.1
1990	5.6	1.0	377.8	23.3
1991	7.5	0.8	360.4	17.8
1992	4.5	0.7	299.0	16.6
1993	2.6	0.4	296.4	15.1
1994	1.3	0.3	291.0	17.0
1995	7.0	0.5	323.5	16.6
1996	4.3	0.3	301.2	17.4
1997	2.4	0.3	225.2	19.5
1998	1.1	0.4	240.7	14.5
1999	8.1	0.5	264.7	14.1
2000	−1.0	0.4	193.6	14.1

Ammonium (kg/yr)		Dissolved silica (kg/yr)	
Residual	Cumulative error	Residual	Cumulative error
62.7	5.5	3693.1	446.5
−11.1	5.1	1997.2	441.4
70.3	6.5	3143.1	392.7
30.7	2.8	3519.6	432.4
18.6	2.6	2853.5	363.0
37.5	3.3	2173.5	434.5
9.5	3.5	2895.5	427.6
26.7	3.3	2744.3	363.2
54.1	4.0	2552.9	372.4
65.1	3.4	3102.4	454.3
60.2	3.6	2777.0	436.5
37.2	3.4	2768.6	409.9
26.6	3.0	2577.5	393.4
47.5	3.0	2092.6	373.5
57.6	3.2	3093.4	335.0
50.4	3.1	3185.6	449.2
50.3	3.2	2379.9	444.4
30.7	2.2	2917.0	394.7
55.4	2.8	3088.0	381.7
13.3	4.0	2486.4	391.2

TABLE 5-II. *Summary of the contribution of solute budget components to the cumulative error in the solute budgets for Mirror Lake, 1981 to 2000*

Solute	Number of years	Precipitation	Tributary streams	Groundwater input	Groundwater output	Surface outlet	Change in storage
Calcium							
Average annual % contribution	20	<0.1	4.0	21.0	68.1	6.7	0.2
Minimum and maximum %		<0.1:<0.1	1.6:8.7	15.1:28.9	44.8:81.7	0.9:32.2	<0.1:1.0
Magnesium							
Average annual % contribution	20	<0.1	4.0	18.7	70.4	6.8	0.2
Minimum and maximum %		<0.1:<0.1	1.5:8.4	12.2:25.6	45.6:84.8	0.9:33.4	<0.1:1.2
Sodium							
Average annual % contribution	20	<0.1	5.0	15.0	73.0	6.8	0.2
Minimum and maximum %		<0.1:<0.1	1.6:12.2	9.6:23.1	44.2:86.3	0.9:32.9	<0.1:0.9
Potassium							
Average annual % contribution	20	<0.1	2.1	38.8	53.7	5.0	0.2
Minimum and maximum %		<0.1:<0.1	0.9:4.4	27.4:48.3	38.3:70.4	0.7:28.2	<0.1:1.7
Hydrogen ion							
Average annual % contribution	20	98.2	0.2	<0.1	1.2	0.3	<0.1
Minimum and maximum %		94.83:99.2	<0.1:0.8	<0.1:<0.1	0.5:3.3	<0.1:1.7	<0.1:0.1

Chloride						
Average annual % contribution	20	<0.1	5.6	1.5	84.5	8.1	0.4
Minimum and maximum %		<0.1:<0.1	1.7:13.6	0.8:2.6	51.0:95.3	1.0:39.1	<0.1:1.8
Sulfate						
Average annual % contribution	20	0.2	6.1	11.9	73.4	7.7	0.6
Minimum and maximum %		0.1:0.3	2.0:14.2	7.3:15.7	48.0:87.0	1.0:36.1	<0.1:3.6
Acid-neutralizing capacity						
Average annual % contribution	9	<0.1	0.9	62.0	35.8	0.7	0.8
Minimum and maximum %		<0.1:<0.1	0.6:1.3	53.2:78.5	16.1:45.3	0.4:1.6	<0.1:2.7
Phosphorus						
Average annual % contribution	20	11.7	3.4	2.6	54.3	6.7	23.9
Minimum and maximum %		0.9:67.2	0.6:10.6	0.1:11.1	25.0:84.0	0.4:42.8	<0.1:66.3
Nitrate						
Average annual % contribution	20	76.3	1.0	<0.1	14.7	5.2	2.8
Minimum and maximum %		43.4:99.1	<0.1:4.5	<0.1:0.1	1.1:35.2	<0.1:27.8	<0.1:13.5

(continued)

TABLE 5-11 (CONTINUED).

Solute	Number of years	Precipitation	Tributary streams	Groundwater input	Groundwater output	Surface outlet	Change in storage
		Ammonium					
Average annual % contribution	20	44.0	1.4	3.0	33.2	5.2	13.1
Minimum and maximum %		6.4:82.8	0.1:6.1	0.8:6.3	7.5:59.7	0.2:23.3	0.1:63.8
		Dissolved silica					
Average annual % contribution	20	<0.1	13.5	60.5	22.1	2.7	1.2
Minimum and maximum %		<0.1:<0.1	6.5:25.3	52.5:76.4	11.5:34.5	0.2:8.8	0.1:3.6
		Dissolved organic carbon					
Average annual % contribution	4	0.2	6.6	1.6	90.2	1.4	<0.1
Minimum and maximum %		0.2:0.2	4.2:9.7	1.2:2.0	87.0:93.4	1.0:1.6	<0.1:<0.1
		Dissolved inorganic carbon					
Average annual % contribution	7	<0.1	1.1	70.9	27.0	0.6	0.3
Minimum and maximum %		<0.1:<0.1	0.9:2.1	65.9:78.6	17.7:32.4	0.:1.5	<0.1:1.5

NOTE: Annual average percent values are rounded to the nearest tenth of a percent.

TABLE 5-12. *Comparison of the residual and the cumulative error of magnesium budgets for Mirror Lake, 1981 to 2000*

Year	Residual of solute budget (kg/yr)	Cumulative error of solute budget; uncertainty in groundwater fluxes used to determine cumulative error[a]			
		50 percent	25 percent	10 percent	5 percent
1981	36.2	100.6	60.1	42.3	39.1
1982	6.5	102.2	53.4	27.0	20.6
1983	49.6	91.9	**49.4**	27.5	22.7
1984	30.6	90.3	49.5	**29.3**	**25.1**
1985	39.7	88.0	45.5	**22.1**	**16.2**
1986	10.0	95.7	51.4	28.5	23.5
1987	−7.2	107.5	55.2	25.9	18.2
1988	−50.0	116.3	58.8	**25.4**	**15.6**
1989	53.3	103.6	55.0	**29.5**	**23.7**
1990	52.9	90.2	**49.0**	28.0	23.4
1991	−13.7	106.6	54.3	24.3	16.0
1992	−3.6	102.1	52.1	23.6	15.7
1993	−50.3	109.0	55.2	**23.9**	**14.7**
1994	−31.7	103.9	52.7	**23.1**	**14.6**
1995	−1.3	93.8	47.7	21.3	13.9
1996	38.3	89.1	47.2	**25.0**	**19.9**
1997	−83.2	105.9	**53.9**	24.0	15.6
1998	32.2	92.2	47.1	**21.6**	**14.6**
1999	6.0	103.5	52.5	23.1	14.7
2000	−26.6	96.1	49.0	**22.0**	**14.6**

NOTE: The cumulative error in the solute budget varies based on the uncertainty in groundwater fluxes used in the calculation of the cumulative error. Values for uncertainty in groundwater fluxes include 25 percent, the value used in all of the budget calculations (chapter 2 and this chapter), and hypothetical values of 50, 10, and 5 percent.

[a] Bold numerals indicate cases where the value of the cumulative error of the solute budget is less than the absolute value of the residual of the solute budget.

TABLE 5-13. *Solute mass in the solute budget of Mirror Lake not attributed to hydrologic solute fluxes, change in solute mass in the lake, or the cumulative error in the solute budget of the lake*

Year	Phosphate (kg/yr)	Nitrate (kg/yr)	Ammonium (kg/yr)	Dissolved silica (kg/yr)	Hydrogen ion (eq/yr)
1981	2.8	363.6	57.1	3246.6	12,203.1
1982	1.9	197.5	−16.2	1555.9	9289.8
1983	1.4	230.4	63.8	2750.4	9403.5
1984	7.7	225.0	28.0	3087.2	9350.6
1985	−11.8	229.7	16.0	2490.4	8960.7
1986	2.7	267.6	34.2	1739.0	9758.1
1987	−11.9	196.1	6.0	2467.9	7750.2
1988	20.8	256.6	23.4	2381.1	9189.1
1989	−6.3	196.1	50.1	2180.5	9929.8
1990	4.6	354.5	61.6	2648.1	12,246.9
1991	6.7	342.7	56.6	2340.5	9956.7
1992	3.8	282.5	33.8	2358.7	7737.7
1993	2.2	281.3	23.6	2184.1	8409.9
1994	1.1	274.0	44.5	1719.1	8945.5
1995	6.5	306.9	54.5	2758.4	8663.1
1996	3.9	283.7	47.4	2736.5	9078.5
1997	2.1	205.7	47.1	1935.5	8220.0
1998	0.7	226.1	28.6	2522.3	7521.9
1999	7.6	250.7	52.7	2706.3	6492.8
2000	−1.4	179.5	9.3	2095.2	7909.9
20-year average	2.0	257.5	36.1	2395.2	9050.89

NOTE: Values are the result of subtracting cumulative error from the budget residual.

TABLE 5-14. *Average annual concentrations in Mirror Lake for the period 1981 to 2000, estimated from chemical model and measured as volume-weighted average*

	Method	N	Mean	Median	Standard deviation	Minimum	Maximum
Calcium	Estimated	20	2.26	2.26	0.11	2.05	2.43
(mg L^{-1})	Measured	20	2.41	2.425	0.11	2.23	2.68
Magnesium	Estimated	20	0.49	0.49	0.02	0.44	0.54
(mg L^{-1})	Measured	20	0.53	0.53	0.02	0.49	0.57
Sodium	Estimated	20	2.07	2.09	0.24	1.70	2.42
(mg L^{-1})	Measured	20	2.22	2.19	0.26	1.86	2.58
Potassium	Estimated	20	0.46	0.45	0.03	0.41	0.52
(mg L^{-1})	Measured	20	0.48	0.47	0.03	0.43	0.55
Chloride	Estimated	20	2.79	2.86	0.45	1.97	3.49
(mg L^{-1})	Measured	20	3.01	3.135	0.45	2.26	3.60
Sulfate	Estimated	20	4.53	4.47	0.44	3.86	5.42
(mg L^{-1})	Measured	20	4.63	4.50	0.51	3.95	5.56
Hydrogen ion	Estimated	10	0.51	0.54	0.14	0.26	0.68
(μeq L^{-1})[a]	Measured	10	0.31	0.305	0.05	0.24	0.40
Phosphate	Estimated	20	0.007	0.007	0.004	0.002	0.016
(mg L^{-1})	Measured	20	0.007	0.006	0.005	0.001	0.020
Nitrate	Estimated	20	0.339	0.376	0.149	0.112	0.580
(mg L^{-1})	Measured	20	0.050	0.044	0.023	0.023	0.103
Ammonia	Estimated	20	0.024	0.023	0.010	0.01	0.046
(mg L^{-1})	Measured	20	0.023	0.0175	0.015	0.01	0.058
Dissolved	Estimated	20	2.26	2.282	0.37	1.56	3.03
silica (mg L^{-1})	Measured	20	2.21	2.205	0.28	1.66	2.80

[a] 1991–2000.

TABLE 5-15. *Comparison of annual water volume, in 1000 m³, determined from the net of precipitation, tributary, surface outlet, evaporation, and change in lake volume and annual water volume determined as the difference between groundwater inputs and groundwater outputs*

Year	Precipitation + tributaries − surface outlet − evaporation − change in lake volume	Ground water input − ground water output
1981	280.9	−194.0
1982	220.1	−230.8
1983	313.9	−189.2
1984	308.2	−183.5
1985	227.8	−180.8
1986	316.6	−208.6
1987	247.6	−264.4
1988	203.7	−292.6
1989	255.4	−247.0
1990	429.3	−193.5
1991	266.3	−260.9
1992	249.8	−252.1
1993	184.1	−281.1
1994	207.0	−259.8
1995	257.8	−227.9
1996	291.8	−197.1
1997	200.7	−287.1
1998	291.4	−245.8
1999	264.9	−282.0
2000	270.8	−248.9

REFERENCES

Andersson-Calles, U.M., and E. Eriksson. 1979. Mass balance of dissolved inorganic substances in three representative basins in Sweden. *Nordic Hydrology* 10:99–114.

Bailey, A.S., J.W. Hornbeck, J.L. Campbell, and C. Eagar. 2003. Hydrometeorological database for Hubbard Brook Experimental Forest: 1955–2000. USDA Forest Service, Northeastern Research Station, General Technical Report NE-305, 36 pp.

Brown, B.E., and D.S. Cherkauer. 1992. Phosphate and carbonate mass balances and their relationships to groundwater inputs at Beaver Lake, Wakesha County, Wisconsin. Tech. Rep. WIS-WRC-9191. Madison: Wisconsin Water Resources Center, 43 pp.

Bukaveckas, P.A., G.E. Likens, T.C. Winter, and D.C. Buso. 1998, A comparison of methods for deriving solute flux rates using long-term data from streams in the Mirror Lake watershed. *Water, Air, and Soil Pollution* 105:277–293.

Buso, D.C., G.E. Likens, and J.S. Eaton. 2000. Chemistry of precipitation, stream water and lake water from the Hubbard Brook Ecosystem Study: A record of sampling protocols and analytical procedures. General Tech. Report NE-275. Newtown Square, Pennsylvania: USDA Forest Service, Northeastern Research Station, 52 pp.

Cogbill, C.V., G.E. Likens, and T.J. Butler. 1984. Uncertainties in historical aspects of acid precipitation: Getting it straight. *Atmospheric Environment* 18(10):2261–2270.

Cole, J.J., N.F. Caraco, and G.E. Likens. 1990. Short-range atmospheric transport: A significant source of phosphorus to an oligotrophic lake. *Limnol. Oceanogr.* 35(6):1230–1237.

Dillon, P.J., and F.H. Rigler. 1974. A test of a simple nutrient budget model predicting the phosphorus concentration in lake water. *Journal of the Canadian Fisheries Research Board of Canada* 31:1771–1778.

Healy, R.W., T.C. Winter, J.W. LaBaugh, and O.L. Franke. 2007. Water budgets: Foundations for effective water-resources and environmental management. U.S. Geological Survey Circular 1308, 90 pp. http://pubs.usgs.gov/circ/2007/1308/

Hem, J.D. 1989. Study and interpretation of the chemical characteristics of natural water, third edition. U.S. Geological Survey Water-Supply Paper 2254. Washington, DC: United States Government Printing Office, 263 pp.

Kelvin, Lord. 1883. Electrical units of measure. *PLA*, vol. 1. http://zapatopi.net/Kelvin/quotes.

Krabbenhoft, D.P., and K.E. Webster. 1995. Transient hydrogeological controls on the chemistry of a seepage lake. *Water Resour. Res.* 31(9):2295–2305.

LaBaugh, J.W. 1985. Uncertainty in phosphorus retention, Williams Fork Reservoir, Colorado. *Water Resour. Res.* 21(11):1684–1692.

LaBaugh, J.W. 1991. Spatial and temporal variation in the chemical characteristics of ground water adjacent to selected lakes and wetlands in the North Central United States. *Verh. Internat. Verein. Limnol.* 24:1588–1594.

LaBaugh, J.W., D.O. Rosenberry, and T.C. Winter. 1995. Groundwater contribution to the water and chemical budgets of Williams Lake, Minnesota, 1980–1991. *Canadian Journal of Fisheries and Aquatic Sciences* 52:754–767.

LaBaugh, J.W., T.C. Winter, D.O. Rosenberry, P.F. Schuster, M.M. Reddy, and G.R. Aiken. 1997. Hydrological and chemical estimates of the water balance of a closed-basin lake in north central Minnesota. *Water Resour. Res.* 33:2799–2812.

Likens, G.E. (ed.). 1985. *An Ecosystem Approach to Aquatic Ecology: Mirror Lake and Its Environment.* New York: Springer-Verlag, 516 pp.

Likens, G.E. 2000. A long-term record of ice cover for Mirror Lake, New Hampshire: Effects of global warming? *Verh. Internat. Verein. Limnol.* 27:2765–2769.

Likens, G.E., and F.H. Bormann. 1995. *Biogeochemistry of a Forested Ecosystem, Second Edition.* New York: Springer-Verlag, 159 pp.

Likens, G.E., D.C. Buso, and J.W. Hornbeck. 2002. Variation in chemistry of stream water and bulk deposition across the Hubbard Brook Valley, New Hampshire, USA. *Verh. Internat. Verein. Limnol.* 28(1):402–409.

Rechow, K.H., S.C. and Chapra. 1979. A note on error analysis for a phosphorus retention model. *Water Resour. Res.* 15(6):1643–1646.

Rosenberry, D.O., P.A. Bukaveckas, D.C. Buso, G.E. Likens, A.M. Shapiro, and T.C. Winter. 1999. Migration of road salt to small New Hampshire lake. *Water, Air, and Soil Pollution* 109:179–206.

Rosenberry, D.O., T.C. Winter, D.C. Buso, and G.E. Likens. 2007. Comparison of 15 evaporation methods applied to a small mountain lake in the northeastern USA. *Journal of Hydrology* 340:149–166.

Sacks, L.A. 2002. Estimating ground-water inflow to lakes in central Florida using the isotope mass-balance approach. U.S. Geological Survey Water Resources Investigations Report 02-4192, 59 pp.

Sacks, L.A., A. Swancar, and T.M. Lee. 1998. Estimating ground-water exchange with lakes using water-budget and chemical mass-balance approaches for ten lakes in ridge areas of Polk and Highlands Counties,

Florida. U.S. Geological Survey Water-Resources Investigations Report, 98-4133, 52 pp.

Scheider, W.A., J.J. Moss, and P.J. Dillon. 1979. Measurement and uses of hydraulic and nutrient budgets. In *Lake Restoration* (pp. 77–83). Minneapolis, Minnesota: U.S. Environmental Protection Agency, Rep. U.S. EPA 440/5-29-001.

Schindler, D.W., R.W. Newbury, K.G. Beaty, and P. Campbell. 1976. Natural water and chemical budgets for a small Precambrian lake basin in central Canada. *Journal of the Fisheries Research Board of Canada* 33(11):2526–2543.

Steel, R.G.D., and J.H. Torrie. 1980. *Principles and Procedures of Statistics: A Biometrical Approach, Second Edition*. New York: McGraw Hill, 613 pp.

Stelzer, R.S., and G.E. Likens. 2006. Effects of sampling frequency on estimates of dissolved silica export by streams: The role of hydrological variability and concentration-discharge relationships. *Water Resour. Res.* 42(7):W07415.

Wentz, D.A., W.J. Rose, and K.E. Webster. 1995. Long-term hydrologic and biogeochemical responses of a soft water seepage lake in north central Wisconsin. *Water Resour. Res.* 31(3): 199–212.

Winter, T.C. 1981. Uncertainties in estimating the water balance of lakes. *Water Resources Bulletin* 17(1):82–115.

Winter, T.C., D.C. Buso, D.O. Rosenberry, G.E. Likens, A.M. Sturrock Jr., and D.P. Mau. 2003. Evaporation determined by the energy-budget method for Mirror Lake, New Hampshire. *American Society of Limnology and Oceanography, Inc.* 48(3):995–1009.

6

MIRROR LAKE: PAST, PRESENT, AND FUTURE

GENE E. LIKENS AND JAMES W. LABAUGH

Mirror Lake is similar in many ways to a vast number of small to medium-sized lakes in the northeastern United States and the Canadian Shield of southeastern Canada. These lakes historically have clear to lightly stained water and are relatively nutrient poor, because of the slowly weathering bedrock of their drainage basins. Recently, these lakes have been affected by numerous environmental impacts, including cultural eutrophication, human development, acid rain, numerous invasive species, and climate change—the combination of which is actually or potentially changing their hydrological and biogeochemical characteristics and temporal trends in a very short time.

Mirror Lake has been studied intensively from the mid-1960s until the present, but 1981–2000 represents a period where water and chemical flux to and from the lake were measured quantitatively, such that no components of these fluxes, such as ground water, were estimated by difference. Thus, the 20-year period covered by this book, 1981–2000, provides a unique snapshot of the hydrology and biogeochemistry of Mirror Lake within its history and natural aging and relative to the various air-land-water fluxes and interactions that have changed as a result of human activities in both the lake's watershed and its airshed.

HISTORICAL CHANGE

It is interesting to reflect on what has happened to Mirror Lake since it was formed from the melting of a large block of ice in the outwash plain of the retreating glacier some 14,000 years ago (Likens and Davis 1975). According to an interpretation of the long-term paleolimnological record (Davis et al. 1985), both the watershed and the lake have undergone marked change since the lake was formed. Much of the following is based upon that interpretation.

During the late glacial period (the first 4000 years), the lake was cold, clear, and nutrient poor and was situated in a tundra-like landscape; transport of freshly eroded materials dominated the chemical inputs to the lake. It is likely that the chemistry of the lake at that time was characterized by $Ca(HCO_3)_2$ (Reynolds and Johnson 1972), based on the weathering of eroded materials. By 10,000 years B.P. the climate had warmed, approximating modern temperatures; an open deciduous forest was developing in the catchment, but the water in the lake was still clear and biologically unproductive. Productivity increased slowly until about 6000 years B.P. and then remained low and relatively constant until the cultural period, some 200 years ago.

Overall, as the climate warmed during the Holocene, more deciduous vegetation developed in the catchment, thereby stabilizing soils and providing tree boles large enough to form organic debris dams in stream channels, and thus reducing the input of eroded materials to the lake (see Likens and Moeller 1985).

There was little or no transpirational water loss until vegetation became established in the catchment. Currently, transpiration in the Mirror Lake basin, not including the lake surface, is about 70 percent of evapotranspiration (Likens and Bormann 1995). This change in transpiration is important because water lost as vapor does not transport solutes. If the water now lost by evapotranspiration were instead lost by runoff, that runoff would remove more solutes than currently occurs. So, with a greater proportion of water lost as runoff prior to development of widespread vegetational cover, the potential for erosion and solute loss from the watershed

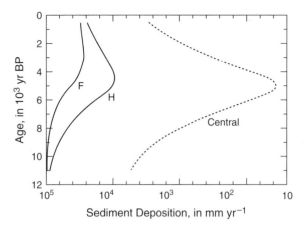

FIGURE 6-1. Estimated historical rates of sedimentation in Mirror Lake, as extrapolated from the central core (Central) to the entire lake basin, using Lehman's frustum (F) and hyperboloid (H) models of deposition (from Davis and Ford 1985).

and transport to the lake would have been higher. Also, there was much less infiltration and more overland flow until soil development occurred during the first 4000 years or so. More erosion would have occurred with more surface flow and less groundwater flow. Sedimentation in the lake peaked about 5000 years B.P. (Fig. 6-1). During cold/less acidic periods (i.e., tundra) with smaller amounts of organic or anthropogenic acids, except for carbonic acid (Reynolds and Johnson 1972), weathering rates would have been smaller. On the other hand, with fresh mineral surfaces and solifluction, there would have been greater potential for weathering release, particularly of P and base cations.

During the first 10,000 years, the basin filled with sediment, reducing the maximum depth from about 24 to 12 m (Davis and Ford 1985). A substantial decline in hemlock trees (*Tsuga canadensis*) in the watershed occurred around 4800 years B.P., possibly due to a pathogen (Davis 1985), and deciduous trees became more abundant. These changes represented major perturbations to the lake's watershed during these first 10,000 years.

Numerous and rapid changes affecting Mirror Lake occurred following European settlement in the watershed about 200 years ago. Like many New England water bodies in the nineteenth century, Mirror Lake was used as a site for intense local industrial and agricultural activity. For example, the water level in Mirror Lake was raised 1 to 2 m by a dam constructed at the outlet; a large fraction of Hubbard Brook was diverted via a 500 m long canal into the lake for an estimated 60 years; the forest along the northern shore was cleared for farming, primarily pasture, during the mid- to late 1800s; a tannery, using large amounts of locally gathered hemlock bark was built at the outlet in the mid-1880s. Erosion probably increased somewhat due to damage to stream channels through clearing and tilling of land, but erosion and transport of particulate matter was relatively small, as the soils in the catchment were not plowed extensively, and the forest regrew rapidly after clearing (Davis et al. 1985; Likens 1985; McLauchlan et al. 2007). There may have been some increased nitrogen loading to the lake following forest clearing and farming. Local industrial and agricultural activity waned after the first decade of the twentieth century, and the lake was largely valued for aesthetics and recreation after that (Likens 1985). Starting in the mid-1950s, external impacts on the hydrological and biogeochemical characteristics of the lake increased substantially due to many factors: acid rain, atmospheric deposition of lead and other heavy metals, increased human settlement around the lake, the construction of an interstate highway through the watershed of the Northeast Tributary, the construction of an access road through the West and Northeast watersheds to the lake, and the impacts of climate change (e.g., there are now about 20 fewer days of ice covering Mirror Lake than in the mid-1960s; see chapter 1, Likens 2000). Clearly, human impacts on the air-land-water interchanges have been larger and more diverse during the last 1 percent of the lake's history than natural factors were in the previous 99 percent of the lake's history. A quantitative analysis of these recent impacts is given in chapter 3.

About 200 years ago, there was undoubtedly increased loading of dissolved inorganic nitrogen from soil disturbance and agriculture; starting about 60 years ago, there would have been increased loading of sulfate and nitrate from acid rain leading to water chemistry characterized by $CaSO_4$. In contrast, some 10,000 to 5000 years ago, Mirror Lake probably was ultra-oligotrophic with water characterized by $(Ca(HCO_3)_2$. About 35 years ago, the lake was impacted by road salt applied to Interstate 93 passing through the Northeast sub-watershed. An estimate of the change

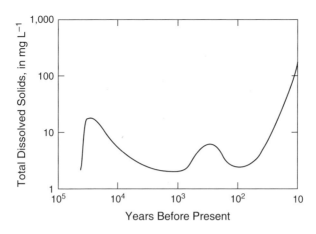

FIGURE 6-2. Estimated concentrations of total dissolved solids in Mirror Lake during the past 14,000 years.

in total dissolved solids in Mirror Lake during the past 14,000 years is given in Fig. 6-2. Currently, the lake is still oligotrophic to slightly mesotrophic (co-limited by nitrogen and phosphorus), and the chemistry is dominated by calcium and sulfate/chloride. Using Barica's modification of the Filatov classification (Barica 1975), the water of Mirror Lake in 2000 would be considered a mixed chloride-bicarbonate, calcium-sodium water type.

HYDROLOGICAL AND BIOGEOCHEMICAL FLUXES

Lakes are connected to their watersheds and airsheds (air-land-water interactions) through hydrological and biogeochemical fluxes, including movement of animals, such as water fowl (Fig. 6-3; Likens and Bormann 1985). The openness of lake ecosystems to these fluxes is a highly visible and vital characteristic for the lake's function and role in the landscape. As such, the trophic condition and biogeochemical status of a lake largely reflect the inputs from the watershed and airshed. Some of the fluxes (e.g., flow of ground water and evaporation) are not readily visible but nevertheless can be extremely important to the mass balance of water and chemicals in a lake. Possibly because such flows are not as visible, they often are not accounted for or measured quantitatively in studies of

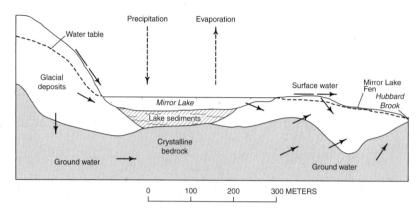

FIGURE 6-3. Hydrologic fluxes into and out of Mirror Lake. All of the hydrologic fluxes can transport dissolved chemicals except evapotranspiration. Larger amounts of eroded particulate matter are transported by higher streamflows.

air-land-water exchanges. As described in the previous chapters in this book, we have endeavored in these long-term studies to measure directly all of the flows into and out of Mirror Lake, and we have presented that accounting for the period 1981–2000 in chapters 2 and 3.

A record of hydrological and chemical characteristics of a lake and its catchment compiled over many years provides the opportunity for analysis of processes revealed by temporal patterns, such as trends in chemical characteristics of atmospheric deposition (Likens and Bormann 1995; Likens et al. 2001, 2005), long-term increases in streamflow related to increases in precipitation (McCabe and Wolock 2002), or loss of vegetation (Likens et al. 1970). Examination of data for New England during the twentieth century indicates that changes in the timing of flow in some rivers are likely due to temperature and precipitation increases (Hodgkins et al. 2003). In such cases, the dates when the bulk of the flow occurs in the spring (snow melt) have been occurring significantly earlier in recent years. Although there are no century-long trends in annual precipitation amount or annual streamflow for the Hubbard Brook Valley (A. Bailey et al. 2003), trends over the past half-century are identifiable; mean annual air temperature (particularly during the winter) has significantly increased, maximum snow depth and snow water content have significantly decreased, and possibly most importantly the duration of snow cover in the experimental watersheds of the Hubbard Brook Valley has decreased by about 23 days during the past 50 years (Campbell et al. 2007).

One of the rivers examined in the Hodgkins et al. (2003) study of climate change was the Pemigewasset River, in New Hampshire. Water that flows from Mirror Lake, through ground water or by the lake's outlet into Hubbard Brook, eventually flows into the Pemigewasset River, reaching the U.S. Geological Survey gauging station on the Pemigewasset River at Plymouth, New Hampshire—some 50 km south of Mirror Lake. The trend at this gauging station was toward an earlier date for the bulk of flow during spring in recent years, but the trend was not statistically significant over the entire record—1904 to 2000—used in the Hodgkins et al. analysis (Hodgkins et al. 2003; also see Cullity 2005). When the more recent period, 1951 to 2000, was examined, the trend for earlier dates for the bulk of flow during spring was statistically significant for the Pemigewasset River at Plymouth. This result correlates with a statistically earlier date for ice out on Mirror Lake (Likens 2000).

The 20-year record for the streams flowing into Mirror Lake (chapter 2) does not clearly indicate that the bulk of streamflow into the lake is taking place at earlier dates in recent years. This result may be a function of the small size of the Mirror Lake watersheds, relative to those of the Pemigewasset River. Furthermore, Hodgkins et al. (2003) found that the New England rivers they examined had no significant trend in flow when the median measured maximum snow depths in their drainage basins were less than 51 cm. Maximum snow depth at snow course Station 2 of the Hubbard Brook Experimental Forest averaged about 60 to 80 cm during a 50-year period that includes the period of study that is the focus of this book (Campbell et al. 2007).

Initial cursory examination of the relation between the El Niño Southern Oscillation (ENSO) and wet years at Mirror Lake is unremarkable, but the maximum amplitude in precipitation between the wettest year and the driest was only about twofold during 1981 to 2000. Recall that in chapter 2 we noted that the 20-year record of interest represents a period of wetter years within the past century. Complex interactions of the ENSO, Pacific Decadal Oscillation, and Atlantic Multidecadal Oscillation influence drought conditions (McCabe et al. 2004). Hydrological conditions at Mirror Lake, however, are not clearly related to those influences.

For example, 1996 was a wet year for Mirror Lake but a year of broad drought across the conterminous United States, based on the McCabe et al. (2004) analysis. Yet the years 1999 to 2002 were years of widespread drought in the conterminous United States, and 1999–2000 was a period at Mirror Lake with little summertime rainfall. In addition, the periods of drought episodes of six years or more (1924–1943 and 1947–1966) in the conterminous United States examined by McCabe et al. (2004) included periods in which the Palmer Hydrologic Drought Index indicated drier than normal climate for the Mirror Lake region (chapter 2, Fig. 2-16). Whereas large-scale phenomena affecting climate may set the stage for general hydrological conditions of a lake and its watershed, the lack of exact correspondence of changing water levels in the lake to such phenomena point out the need to keep in mind the local climate influences on the hydrological characteristics of a lake. In addition, there is a time lag between when water enters the groundwater flow system and when that water discharges to the lake. In the case of Mirror Lake, an important characteristic is its relation to ground water, which provides fairly consistent flows of water and chemicals into the lake, as well as serving as a conduit through which water and chemicals depart in a consistent and predictable way (see chapter 3). For Mirror Lake, 59 percent of the water for the lake came from surface inlet flow, and 51 percent of the outflow from the lake occurred as groundwater flow (see chapter 2).

MANAGEMENT CONSIDERATIONS

Scientific guidance of public policy is not necessarily a straightforward process (e.g., Likens 1992; Lackey 2007). Nevertheless, the long-term scientific information and understanding that we have generated for Mirror Lake is potentially of great value for resource managers and decision makers. For example, information regarding eutrophication, acid rain, and road-salt effects has been used to guide the management of Mirror Lake and other lakes in the northeastern United States and southeastern Canada.

DISSOLVED INORGANIC NITROGEN

Prior to the cultural period, input of dissolved inorganic nitrogen to the lake and to the watersheds of the lake came from two sources: atmospheric deposition and N-fixation, as there are negligible amounts of weathering substrates bearing N in the Hubbard Brook Valley. Biotic fixation of atmospheric N in the Hubbard Brook Valley is thought to be very small and localized (Roskoski 1980; Bormann et al. 1977; Steinhart et al. 2000). Thus, it can be assumed that the nitrogen in atmospheric deposition was the primary source for Mirror Lake in the past. Once added to the catchment and lake, however, the cycling is efficient, and net retention is high (chapters 3 and 5). McLauchlan et al. (2007, p. 7466) have suggested that nitrogen availability, defined as the "supply of N to terrestrial plants and soil microorganisms relative to their N demands" has declined during the past 75 years in the Mirror Lake watershed, and that long-past human disturbances such as logging and agriculture affect nitrogen cycling and flux even a century after these disturbances occurred.

Over the past 10,000 years, sediment deposition of nitrogen, phosphorus, and organic matter in Mirror Lake peaked around 6000 to 4000 years B.P. and generally declined thereafter (Fig. 6-4). Part of this change

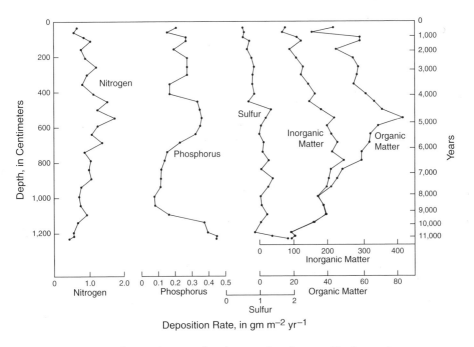

FIGURE 6-4. Sedimentation rates for nitrogen, phosphorus, sulfur, inorganic matter, and organic matter in Mirror Lake during the past 14,000 years (from Likens and Moeller 1985).

may have been related to the sudden decline of hemlock in the watershed (Davis 1985).

Currently, the flux of dissolved inorganic nitrogen into Mirror Lake is dominated by human activity (emissions of NOx to the atmosphere and ammonium and nitrate in septic drainage and fertilizer) and totals about 117 kg N/yr, or 7 to 23 times greater than the Pre–Industrial Revolution (PIR) value (estimated to be between 5 and 18 kg N/yr).

Concentrations of sulfate in the lake have been declining since 1965, and particularly since 1970, whereas dissolved inorganic nitrogen concentrations have been approximately the same since 1964. These trends are reflected in total inputs to the lake as well (Fig. 6-5). Assuming the linear trends observed since the mid-1960s continue, and based on these trends, we might expect that concentrations of sulfate and dissolved

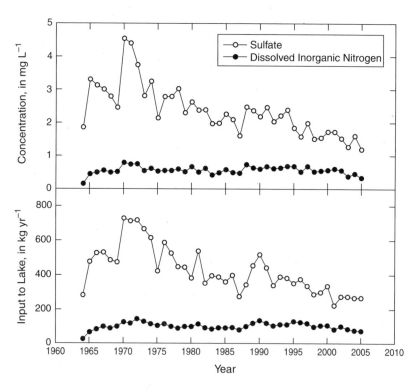

FIGURE 6-5. Concentration and inputs of sulfate and dissolved inorganic nitrogen for Mirror Lake from 1964 to 2005.

inorganic nitrogen in the lake might decline to about 1.0 and 0.03 mg/L, respectively, in 2020. The 20-year (1981–2000) quantitative snapshot provides an exceptional opportunity to place the current flux of critical nutrients, such as phosphate, dissolved inorganic nitrogen, and dissolved silica, for the lake into perspective during the history of the lake.

PHOSPHORUS

Phosphorus occurs, but in low concentrations, within weathering substrates in the Valley (Likens and Bormann 1995; Yanai 1990). Both phosphorus and nitrogen are rapidly taken up by organisms in streams, thereby reducing the fluvial flux (Meyer and Likens 1979; Meyer et al. 1981; Bernhardt et al. 2003, 2005). Phosphorus was presumably much more readily available for weathering release from apatite early in the lake's history. Because the mineral apatite was easily and rapidly weathered from exposed rock faces it became less readily available with time, and now phosphorus release from apatite is small, but its release may be facilitated by mycorrhizal mining of apatite crystals in the interior of rock surfaces (Jongmans et al. 1997; Blum et al. 2002).

With cultural development phosphorus flux probably increased from new anthropogenic sources (e.g., fertilizer, animal manure, and septic drainage from human habitation in the catchment of Mirror Lake), although amounts are relatively small (chapter 3; Figs. 3-5 and 3-6).

Since 1970, biological primary production in Mirror Lake has been co-limited by nitrogen and phosphorus (Gerhart and Likens 1975; Bade et al. 2008). Even with slightly increased inputs of both dissolved inorganic nitrogen and phosphorus since the late 1700s from human activities (Figs. 6-4 and 6-5), the lake can still be considered to be nutrient poor and to have relatively low biological productivity (Bade et al. 2008).

SILICA

In contrast to nitrogen after glacial retreat, the flux of dissolved silica in stream water originated from the weathering of siliceous minerals in the catchment (see inorganic matter in Fig. 6-4). As of 2000, however, a significant source of dissolved silica in drainage water has probably originated from the weathering of biologically generated amorphous silica in the upper soil horizons of the catchment (Saccone et al. 2008). It is striking to note that current concentrations of dissolved silica in stream water flowing into the lake are 2.9 to 5.0 times higher than the concentration in lake water (chapter 3). The difference results from rapid uptake of dissolved silica by diatoms, primarily near the mouths of the inlet tributaries.

CULTURAL EUTROPHICATION

When biogeochemical fluxes are modified by human activity, the trophic state of a lake may change dramatically (e.g., Likens 1972; Schindler 1977; Carpenter et al. 1998; Smith 1998). Traditionally, a major source of N and P to Mirror Lake and its watershed was from the atmosphere via direct precipitation (chapter 3). For example, in the case of Mirror Lake, from 1964 to 1985, 54 percent of the loading of inorganic N and 3 percent of the inorganic P entered the lake through direct precipitation to the lake surface (see Bormann and Likens 1985). From 1981 to 2000, the major source of N and P to both Mirror Lake and its watershed was from the atmosphere via precipitation (chapter 3). In 1985, Bormann and Likens predicted that, with increased human activity in the lake's watershed, runoff sources of P and N would soon dominate. This change did not occur. With increased human habitation around the lake and decreased acid rain (N loading) during the past two decades, 87 percent of inorganic N and 55 percent of inorganic P loading came from the atmosphere. Even with slightly increased inputs of both dissolved inorganic nitrogen and phosphorus from human activities since the late 1700s (Figs. 6-4 and 6-5), the lake is still nutrient poor and has relatively low biological productivity (Bade et al. 2009).

Eutrophication of the lake is minimal, with little change with time, and the lake remains co-limited by N and P as it was 30 years ago (Bade et al. 2008; Gerhart and Likens 1975). Based on the pioneering work by Vollenweider (1968), and as updated (e.g., Vollenweider 1990), the extremely small loading of phosphorus (2.4 mg P/m3) would place Mirror Lake well within the oligotrophic category.

ACID RAIN

After peaking in the mid-1970s, atmospheric deposition of sulfur decreased due to federal regulation, which reduced emissions of SO2 from the source area for Mirror Lake (Likens et al. 2001, 2005). Emissions of NOx increased until about 2000 but have since decreased; likewise, nitrate concentrations in precipitation at Mirror Lake decreased by about 50 percent from 2001 to 2007. Because of the continuing long-term data on Mirror Lake, we will be able to follow, study, and interpret important temporal trends like this into the future.

We documented in chapter 3 the changes observed in chemical concentrations of Mirror Lake during the 20-year period that is the focus of this book. The decline in calcium concentration in Mirror Lake between 1981 and 2000 is consistent with declines observed in headwater streams of the Hubbard Brook Experimental Forest (Fig. 6-6; Likens et al. 1996, 1998) and in other lakes in eastern North America (Jeziorski et al. 2008; and particularly Keller et al. 2001). The decline is associated with a decrease in exchangeable calcium in the soils of these systems, and is a response to the impact of acid rain (e.g., Likens et al. 1996, 1998; Houle et al. 2006). Similar declines are found elsewhere in forest watersheds in eastern North America and Europe (Lawrence et al. 1995; Bailey et al. 2005; Watmough et al. 2005; Warby et al. 2009). The general decline in volume-weighted concentrations of magnesium and potassium in Mirror Lake is similarly consistent with long-term observations for stream water in the Hubbard Brook Experimental Forest watersheds (Fig. 6-6), and in other comparable watersheds.

Based on the difference between inputs and outputs, 9 of the 20 years of study were years in which Mirror Lake lost more calcium in a year than it received, this imbalance being primarily a function of water fluxes (chapter 3). On average, retention of calcium is 2 percent of input (chapter 5). Overall, concentrations and relative importance on a mass and equivalence basis of calcium, magnesium, potassium, and sulfate decreased, whereas concentrations of sodium and chloride increased in Mirror Lake during 1981–2000 (Figs. 6-6, 6-7, and 6-8). These changes tended to balance, causing total cations and total anions to remain approximately the same on an annual basis during this period. The 20-year annual average of total cations was 0.274 milliequivalents per liter with a standard deviation of 0.006. The 20-year annual average of total anions was 0.269 milliequivalents per liter with a standard deviation of 0.006.

The temporal trends for decreasing concentrations of calcium, magnesium, potassium, sulfate, and acidity in Mirror Lake are similar to those in

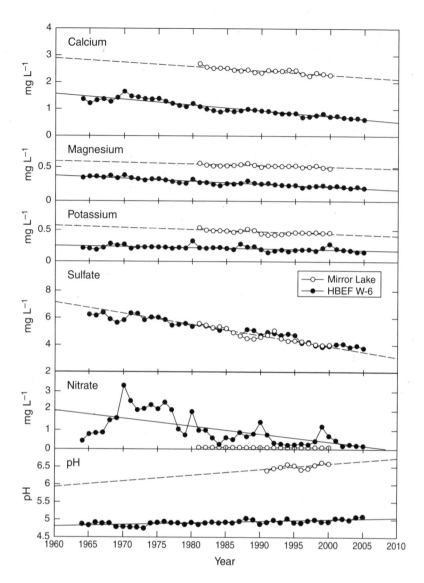

FIGURE 6-6. Comparison of concentrations of calcium, magnesium, potassium, sulfate, nitrate, and pH in Mirror Lake and Watershed 6 (W6) of the Hubbard Brook Experimental Forest from 1964 through 2005.

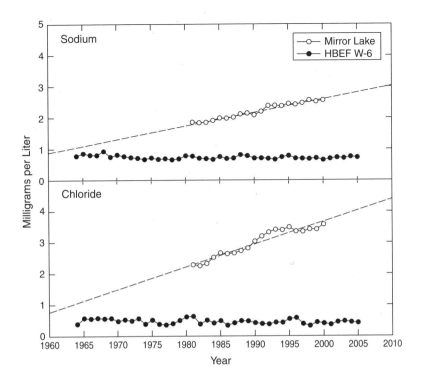

FIGURE 6-7. Comparison of concentrations of sodium and chloride in Mirror Lake and Watershed 6 (W6) of the Hubbard Brook Experimental Forest from 1964 through 2005.

headwater streams of the Hubbard Brook Experimental Forest, whereas the increasing concentrations of sodium and chloride in Mirror Lake contrasted with stable concentrations in headwater streams (Figs. 6-6 and 6-7). It is interesting to note that because of the recent marked decline of streamwater concentrations of nitrate, these concentrations are now similar to those in Mirror Lake (Fig. 6-6).

The relation of highway runoff to biological communities has been of interest for some time (e.g., Buckler and Granato 1999; Forman et al. 2003). Substantial changes have occurred in the physical characteristics of some lakes due to receipt of large quantities of road salt resulting in permanent stratification, or meromixis (Bubeck and Burton 1989). In the case of Mirror Lake, even when physical barriers were put in place to minimize transport of solutes from road-salt application within the Northeast sub-watershed, such solutes still followed hydrologic pathways

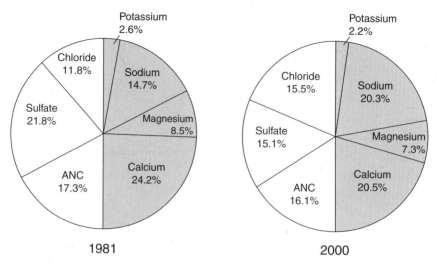

FIGURE 6-8. Chemical composition of Mirror Lake water as percent in 1981 and in 2000.

and entered the lake (Rosenberry et al. 1999). As of 2000, this introduction of sodium and chloride into the lake has had the unexpected effect of maintaining constancy in the total dissolved solids concentration in the lake over a 20-year period in spite of a general decline in the other total dissolved solids input to the lake (Fig. 6-9).

THE FUTURE

Just as the long-term record of water and solutes in precipitation and the headwater streams of the Hubbard Brook Valley led to documentation of acid rain in North America, the 20 years of record in this book reveal another phenomenon that may be important in the years ahead for lakes in similar climatic and geologic settings. That phenomenon is the change in solute abundance due to the combination of declines in calcium and increases in sodium and chloride, whereby the ionic strength and total dissolved solids do not continue to decline as expected, due to changes in solutes in precipitation/watershed processes that are a byproduct of acid deposition, but instead stay the same or eventually increase. What this change means in dilute aquatic ecosystems has yet to be revealed.

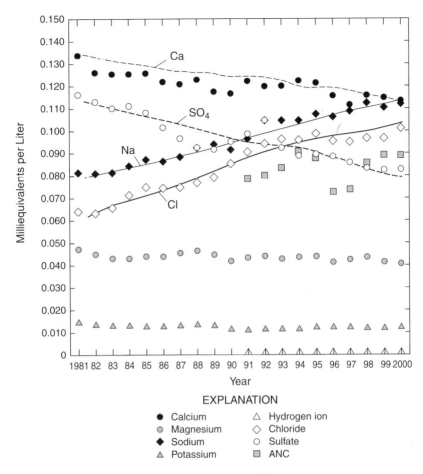

FIGURE 6-9. Changes in annual volume-weighted concentration in milliequivalents per liter of calcium, magnesium, potassium, sodium, sulfate, chloride, hydrogen ion, and acid neutralizing capacity (ANC) in Mirror Lake from 1981 through 2000.

Could continuing inputs of NaCl from road deicers eventually lead to increased acidification or delayed recovery from acidification (see Hindar et al. 1995) of the lake?

What we have seen in that regard at Mirror Lake may appear in lakes elsewhere in northeastern North America, whereby the total dissolved solids in some lakes will not decline as calcium in their watersheds declines due to input of road salt. Yet, in other lakes, those that do not receive input of road salt, total dissolved solids will continue to decline as the sources of calcium and sulfur decline. In such systems, the ecosystem response as total dissolved solids decline, in contrast to the response in aquatic ecosystems where the total dissolved solids do not decline but change in relative abundance of ions, may write a new chapter in the unintended consequences of human activity on the environment.

Most lakes in North America, similar to Mirror Lake, are increasingly subjected to anthropogenic impact and change, as discussed briefly above and more fully in chapter 3. Major impacts, such as shoreline development and degradation (resulting in loss of habitat for fish and other organisms, increased erosion, and input of toxic metals and pesticides), agricultural activity within the catchment (leading to increased erosion and nutrient and pesticide runoff), invasion of non-native species, acid rain, and climate change, are now common threats to the integrity of the structure and function of lake ecosystems worldwide and have become the subject of much research and planning (Myers et al. 2007). Obviously, as is clear from the long-term record for Mirror Lake, that major changes have occurred "naturally" in the lake and its catchment (i.e., vegetation development, basin filling) since the lake was formed 14,000 years ago. These changes are ongoing, as is typical of any ecosystem, because an ecosystem contains living organisms, which change due to biotic succession, evolution, and ecosystem development with time (Likens 1992, 2008). So, Mirror Lake is continuing along an aging pathway, and ultimately, it will fill with sediment and no longer be a lake (Likens 1972).

Some previous, anthropogenic disturbances to Mirror Lake were large, such as forest clearing and diversion of Hubbard Brook through the lake, but the current rate of impact driven by human activities is unprecedented in the history of Mirror Lake: that is, at present, the lake and its watershed are being subjected simultaneously to climate change, acid rain, invasive species, salinization, and shoreline development. For example, one of the dominant tree species in the watershed, sugar maple (*Acer saccharum*) has been present in the watershed for some 10,000 years, peaking in abundance about 3000 to 6000 years ago (Davis 1985). Now, however, the sugar maple is in sharp decline because of impacts of acid rain, primarily from base cation depletion in soils (e.g., Likens et al. 1996; Siccama et al. 2007). Future changes in species composition (e.g., if sugar maple were to be replaced by oak [*Quercus* spp.] and hickory [*Carya spp.*] as climate warms during the next 75 years or so) would have significant impact on biogeochemical flux and cycling; as different species have different effects on nitrogen cycling (Lovett et al. 2004; Lovett and Mitchell 2004; Templer et al. 2005; Union of Concerned Scientists 2006).

Long-term studies—and their vital component, monitoring—are frequently maligned as "mindless data collecting" (Likens 1992; Lovett et al. 2007; Lindenmeyer and Likens 2009) or as "data rich but information poor" (Ward et al. 1986). We would argue strongly, however, that given the complex mix of anthropogenic pressures on the lake, our long-term quantitative perspective on change is unusually valuable to science and to decision makers for several purposes including the following:

- identifying and characterizing emerging environmental problems, such as climate change and salinization,
- identifying temporal trends and extremes, such as wet and dry years or warm winters,
- providing a baseline for judging change in lake ecosystem structure, such as lakewater clarity or species diversity,
- judging change in lake ecosystem function, such as nutrient limitation for lake biological productivity, and
- providing insight and guidance to policy makers and resource managers dealing with the diminishment of ecosystem services.

Furthermore, documenting changes over time, also known as doing mensurative experiments (Hurlbert 1984), has revealed important processes affecting watersheds and the globe—acid rain (Likens et al. 1967; Fisher et al. 1968; Likens et al. 1970), climate variability in relation to timing and magnitude of stream flow (Hodgkins et al. 2003), and the rise in atmospheric carbon (Keeling et al. 1976).

Because our completely measured fluxes of water and chemicals to the lake provide values of input and output with relatively low error (chapters 4 and 5), they can be used by resource managers and policy makers with more confidence and understanding. Such quantitative information can be used to guide pollution control or the development of alternative, less polluting systems, as has been done with sources of acid rain, control of salt contamination from Interstate 93 (chapter 3), and less damaging development of shoreline areas. Thus, these data are important not only for protecting the integrity of Mirror Lake but also for providing insight and guidance regarding the protection and/or management of other surface water bodies similarly affected.

Mirror Lake is affected by the same suite of environmental problems as are other lakes in the northeastern United States and southeastern Canada. Cultural development, including shoreline degradation, is one of the most serious environmental threats to lakes like Mirror Lake at the present time. Nevertheless, regional climate change is the major environmental issue on the horizon for this area, with initial and expected changes in amount and variability of temperature and precipitation (Campbell et al. 2007; Hayhoe et al. 2007) and in species composition in Mirror Lake and its airshed and watershed. Indeed, as mentioned, the ice cover on Mirror Lake already is about 20 days less than compared with the early 1960s (Likens 2000), and this is a widespread phenomenon (e.g., Magnuson et al. 2000). The Union of Concerned Scientists (2006) has predicted that the climate of the area, including the Hubbard Brook Valley, may be similar to Virginia's climate by 2070. The full environmental impacts of these unprecedented changes are unknown but represent many complicated scientific and resource management challenges for the future. Our 20-year snapshot of hydrologic and biogeochemical flux during 1981–2000 will serve us well in evaluating and putting into perspective such conditions experienced in the future.

Former Forest Service Project Leader Robert S. Pierce, G. E. Likens, and the Hubbard Brook Research Foundation have been active and innovative in trying to protect the shoreline of Mirror Lake from further development. It could be argued that a major reason for the maintenance of high-quality (clear, nutrient-poor) water in Mirror Lake resulted from these aggressive activities since the 1980s to protect, through purchase and/or transfer to the U.S. Forest Service, some 70 percent of the shoreline from additional human development and degradation.

Hydrologic and biogeochemical flux represent the critical operative connections for this lake with the atmosphere and the landscape, and indeed with the remainder of the biosphere (Likens and Bormann 1985). Moreover, and possibly even more important relative to current environmental problems, is that these fluxes are the major "pressure points" where management can be applied to protect the structure and function of a lake. Air-land-water connections and interactions then represent the key factors for understanding how a lake "works," for evaluating its abiotic and biotic structure, and for protecting the lake against unprecedented unnatural change.

Although many of the anthropogenic stresses on Mirror Lake are now being better controlled, important stressors continue to pose long-term problems for the health of the lake. Much of the shoreline and catchment are now protected, inputs of acid rain are declining, and structures have been emplaced along highway I-93 to reduce road salt entering the Northeast Tributary to the Lake. Nevertheless, acid rain continues and road salt is still applied to I-93 and other roads in the area, such as the access road to Mirror Lake. Movement of water in the groundwater system flowing into the lake, such as in the vicinity of the Northeast Tributary, is such that water originating from the vicinity of highway I-93, before control structures were emplaced, has yet to discharge to the lake. Moreover, road salt contamination may continue to increase into the future because of increased climate variability, as there may be increased application of salt to deal with more icy conditions. Warmer conditions in winter may lead to the occurrence of precipitation that falls in the form of sleet or freezing rain instead of snow. Changes in the future for Mirror Lake are uncertain with respect to climate change. A change in climate, however, could overwhelm the resistance of the lake for major change-in-state by greatly increasing or decreasing the amount and timing of precipitation, evapotranspiration, flooding, and erosion, by dramatically warming the water, and by fostering a change in the composition and diversity of terrestrial and aquatic species. Given the resilience of the lake during the past 14,000 years, however, it probably is reasonable to assume that the current general characteristics of the lake will persist for at least the next 50 years, even though it is likely that human-accelerated environmental change will be more rapid in the future (Likens 1991). Overall, in the short-term, the decline in acid rain, the structural barrier to movement of road salt to the lake from I-93, and the protection of the majority of shoreline bode well for the lake to maintain its clear, nutrient-poor status into the future.

Today, Mirror Lake exists as a beautiful "gem" in the landscape of the White Mountains. Its functional integrity provides recreation and other highly valued ecosystem services for inhabitants and visitors of the region, as well as a healthy habitat for the organisms that call Mirror Lake home.

As acclaimed Ojibwe writer Louise Erdrich has lamented about parts of the upper Midwest in *Books and Islands in Ojibwe Country,* "I stop at a spot just off the highway, one of those square tubes of rooms facing the road. The line of identical brown doors and windows, like staring faces, has a sullen aspect. No skylarks. The textured siding is a defeated looking tan color. . . . The yard is dust, struggling weeds, trampled gravel. . . . I'm disoriented, as one always is leaving some wild place on the Earth and returning to human disorder. The unattractive nature of the towns and buildings seems purposeful. There is a belligerent streak to the ugliness." One wonders if this could happen at Mirror Lake without diligence to protect the integrity of the lake and its catchment and airshed.

REFERENCES

Bade, D.L., K. Bouchard, and G.E. Likens. 2009. Algal co-limitation by nitrogen and phosphorus persists after 30 years in Mirror Lake (New Hampshire, USA). *Verh. Internat. Verein. Limnol.* 30 (7):1121–1123.

Bailey, A.S., J.W. Hornbeck, J.L. Campbell, and C. Eagar. 2003. Hydrometeorological database for Hubbard Brook Experimental Forest: 1955–2000. USDA Forest Service, Northeastern Research Station, Gen. Tech. Report NE-305, 36 pp.

Bailey, S.W., S.B. Horsley, and R.P. Long. 2005. Thirty years of change in forest soils of the Allegheny Plateau, Pennsylvania. *Soil. Sci. Soc. Am. J.* 69:681–690.

Barica, J. 1975. Geochemistry and nutrient regime of saline eutrophic lakes in the Erickson-Elphinstone district of Southwestern Manitoba. Environment Canada, Fisheries and Marine Service Technical Report no. 511, 82 pp.

Bernhardt, E.S., G.E. Likens, D.C. Buso, and C.T. Driscoll. 2003. In-stream uptake dampens effects of major forest disturbance on watershed nitrogen export. *Proceedings of National Academy of Sciences* 100(18):10304–10308.

Bernhardt, E.S., G.E. Likens, R.O. Hall Jr., D.C. Buso, S.G. Fisher, T.M. Burton, J.L. Meyer, W.H. McDowell, M.S. Mayer, W.B. Bowden, S.E. G. Findlay, K.H. Macneale, R.S. Stelzer, and W.H. Lowe. 2005. Can't see the forest for the stream? In-stream processing and terrestrial nitrogen exports. *BioScience* 55(3):219–230.

Blum, J.D., A. Klaue, C.A. Nezat, C.T. Driscoll, C.E. Johnson, T.G. Siccama, C. Eagar, T.J. Fahey, and G.E. Likens. 2002. Mycorrhizal weathering of apatite as an important calcium source in base-poor forest ecosystems. *Nature* 417:729–731.

Bormann, F.H., and G.E. Likens. 1985. Air and watershed management and the aquatic ecosystem. pp. 436-444. In G. E. Likens (ed.). *An Ecosystem Approach to Aquatic Ecology: Mirror Lake and Its Environment* (pp. 436444). New York: Springer-Verlag.

Bormann, F.H., G.E. Likens, and J.M. Melillo. 1977. Nitrogen budget for an aggrading northern hardwood forest ecosystem. *Science* 196(4293):981–983.

Bubeck, R.C., and R.S. Burton. 1989. Changes in chloride concentrations, mixing patterns, and stratification characteristics of Irondequoit Bay, Monroe County, New York, after decreased use of road-deicing salts, 1974–1984. U.S. Geological Survey Water-Resources Investigations Report 87-4223, 52 pp.

Buckler, D.R., and G.E. Granato. 1999. Assessing biological effects from highway-runoff constituents. U.S. Geological Survey Open-File Report 99-240, 53 pp.

Campbell, J.L., C.T. Driscoll, C. Eagar, G.E. Likens, T.G. Siccama, C.E. Johnson, T.J. Fahey, S.P. Hamburg, R.T. Holmes, A.S. Bailey, and D.C. Buso. 2007. Long-term trends from ecosystem research at the Hubbard Brook Experimental Forest. USDA Forest Service, Northern Research Station, Newtown Square, Pennsylvania. Gen. Tech. Report NRS-7, 41 pp.

Carpenter, S.R., N. Caraco, D.L. Correll, R.W. Howarth, A.N. Sharpley, and V.H. Smith. 1998. Nonpoint pollution of surface waters with phosphorus and nitrogen. *Ecol. Appl.* 8:559–568.

Cullity, A.K. 2005. Trends in New England hydroclimate as reflected in the record of the Pemigewasset River at Plymouth, NH. M.S. Thesis, University of New Hampshire. 87 pp.

Davis, M.B. 1985. History of the vegetation on the Mirror Lake watershed. In G.E. Likens (ed.). *An Ecosystem Approach to Aquatic Ecology: Mirror Lake and Its Environment* (pp. 53–65). New York: Springer-Verlag.

Davis, M.B., and J. Ford. 1985. Late-glacial and Holocene sedimentation. In G.E. Likens (ed.). *An Ecosystem Approach to Aquatic Ecology: Mirror Lake and Its Environment* (pp. 345–355). New York: Springer-Verlag.

Davis, M.B., R.E. Moeller, G.E. Likens, J. Ford, J. Sherman, and C. Goulden. 1985. Paleoecology of Mirror Lake and its watershed. In G. E. Likens (ed.). *An Ecosystem Approach to Aquatic Ecology: Mirror Lake and Its Environment* (pp. 410–429). New York: Springer-Verlag.

Erdrich, L. 2003. *Books and Islands in Ojibwe Country.* Washington, DC: National Geographic Society.

Fisher, D.W., A.W. Gambell, G.E. Likens, and F.H. Bormann. 1968. Atmospheric contributions to water quality of streams in the Hubbard Brook Experimental Forest, New Hampshire. *Water Resour. Res.* 4(5):1115–1126.

Forman, R.T.T., D. Sperling, J.A. Bissonette, A.P. Clevenger, C.D. Cutshall, V.H. Dale, L. Fahrig, R. France, C.R. Goldman, K. Heanue, J.A. Jones, F.J. Swanson, T. Turrentine, and T.C. Winter. 2003. *Road Ecology: Science and Solutions.* Washington, DC: Island Press, 481 pp.

Gerhart, D.Z., and G.E. Likens. 1975. Enrichment experiments for determining nutrient limitation: Four methods compared. *Limnol. Oceanogr.* 20(4):649–653.

Hayhoe, K., C.P. Wake, T.G. Huntington, L. Luo, M.D. Schwartz, J. Sheffield, E. Wood, B. Anderson, J. Bradbury, A. DeGaetano, T. Troy, and D. Wolfe. 2007. Past and future changes in climate and hydrological indicators in the U.S. Northeast. *Clim. Dyn.* 28:381–407.

Hodgkins, G.A., R.W. Dudley, and T.G. Huntington. 2003. Changes in the timing of high river flows in New England over the 20th century. *J. Hydrology* 278:244–252.

Houle, D., R. Ouimet, S. Couture, and C. Gagnon. 2006. Base cation reservoirs in soil control the buffering capacity of lakes in forested catchments. *Can. J. Fish. Aquatic Sci.* 63:471–474.

Hurlbert, S.H. 1984. Pseudoreplication and the design of ecological field experiments. *Ecological Monographs* 54(2):187–211.

Jeziorski, A., N.D. Yan, A.M. Paterson, A.M. DeSellas, M.A. Turner, D.S. Jeffries, B. Keller, R.C. Weeber, D.K. McNicol, M.E. Palmer, K. McIver, K. Arseneau, B.K. Ginn, B.F. Cumming, and J.P. Smol. 2008. The widespread threat of calcium decline in fresh waters. *Science* 322:1374–1377.

Johnson, N.M., R.C. Reynolds, and G.E. Likens. 1972. Atmospheric sulfur: Its effect on the chemical weathering of New England. *Science* 177(4048):514–516.

Jongmans, A.G., N. van Breemen, U. Lundström, P.A.W. van Hees, R.D. Finlay, M. Srinivasan, T. Unestam, R. Giesler, P.-A. Melkerud, and M. Olsson. 1997. Rock-eating fungi. *Nature* 389:682–683.

Keeling, C.D., R.B. Bacastow, A.E. Bainbridge, C.A. Ekdahl Jr., P.R. Guenther, L.S. Waterman, and J.F.S. Chin. 1976. Atmospheric carbon dioxide variations at Mauna Loa Observatory, Hawaii. *Tellus* 28:538–551.

Keller, W., S.S. Dixit, and J. Heneberr. 2001. Calcium declines in northeastern Ontario lakes. Can. J. Fish. *Aquatic Sci.* 58:2011–2020.

Lackey, R.T. 2007. Science, scientist, and policy advocacy. *Conserv. Biol.* 21(1):12–17.

Lawrence, G.B., M.B. David, and W.C. Shortle. 1995. A new mechanism for calcium loss in forest-floor soils. *Nature* 378:162–165.

Likens, G.E. 1972. Eutrophication and aquatic ecosystems. In G. E. Likens (ed.). *Nutrients and Eutrophication* (pp. 3–13). Proceedings of the American Society of Limnology and Oceanography Special Symposia, Vol. 1. Lawrence, Kansas.

Likens, G.E. 1991. Human-accelerated environmental change. *Bioscience* 41(3):130.

Likens, G.E. 1985. Mirror Lake: Cultural history. In G. E. Likens (ed.). *An Ecosystem Approach to Aquatic Ecology: Mirror Lake and Its Environment* (pp. 72–83). New York: Springer-Verlag.

Likens, G.E. 1992. The ecosystem approach: Its use and abuse. Excellence in Ecology, Vol. 3. Oldendorf/Luhe, Germany: Ecology Institute, 167 pp.

Likens, G.E. 2000. A long-term record of ice cover for Mirror Lake, New Hampshire: Effects of global warming? *Verh. Internat. Verein. Limnol.* 27(5):2765–2769.

Likens, G.E. 2008. Lee Talbot's ecosystem approach to conservation: Can it be sustained? In R. Stewart, R.B. Jonas, L.L. Rockwood, and T. Dietz (eds.). *Pathways to Sustainability: A Prospectus* [Lee M. Talbot Festschrift] (pp. 51–61). Fairfax, VA: George Mason University.,

Likens, G.E., and F.H. Bormann. 1985. An ecosystem approach. In G.E. Likens (ed.). *An Ecosystem Approach to Aquatic Ecology: Mirror Lake and Its Environment* (pp. 1–8). New York: Springer-Verlag.

Likens, G.E., and F.H. Bormann. 1995. *Biogeochemistry of a Forested Ecosystem, Second Edition.* New York: Springer-Verlag, 159 pp.

Likens, G.E., F.H. Bormann, N.M. Johnson, D.W. Fisher, and R.S. Pierce. 1970. Effects of forest cutting and herbicide treatment on nutrient budgets in the Hubbard Brook watershed-ecosystem. *Ecol. Monogr.* 40(1):23–47.

Likens, G.E., F.H. Bormann, N.M. Johnson, and R.S. Pierce. 1967. The calcium, magnesium, potassium, and sodium budgets for a small forested ecosystem. *Ecology* 48(5):772–785.

Likens, G.E., and D.C. Buso. 2008. Long-term changes in streamwater chemistry following disturbance in the Hubbard Brook Experimental Forest, USA. *Verh. Internat. Verein. Limnol.* 30 (in press).

Likens, G.E., D.C. Buso, and T.J. Butler. 2005. Long-term relationships between SO2 and NOX emissions and SO_4^{2-} and NO_3^- concentration in bulk deposition at the Hubbard Brook Experimental Forest, New Hampshire. *J. Environ. Monitoring* 7(10):964–968.

Likens, G.E., T.J. Butler, and D.C. Buso. 2001. Long- and short-term changes in sulfate deposition: Effects of the 1990 Clean Air Act amendments. *Biogeochemistry* 52(1):1–11.

Likens, G.E., and M.B. Davis. 1975. Post-glacial history of Mirror Lake and its watershed in New Hampshire, USA: An initial report. *Verh. Internat. Verein. Limnol.* 19(2):982–993.

Likens, G.E., C.T. Driscoll, and D.C. Buso. 1996. Long-term effects of acid rain: Response and recovery of a forest ecosystem. *Science* 272:244–246.

Likens, G.E., C.T. Driscoll, D.C. Buso, T.G. Siccama, C.E. Johnson, G.M. Lovett, T.J. Fahey, W.A. Reiners, D.F. Ryan, C.W. Martin, and S.W. Bailey. 1998. The biogeochemistry of calcium at Hubbard Brook. *Biogeochemistry* 41(2):89–173.

Likens, G.E., and R.E. Moeller. 1985. Chemistry. In G.E. Likens (ed.). An Ecosystem Approach to Aquatic Ecology: Mirror Lake and Its Environment (pp. 392–410). New York: Springer-Verlag.

Lindenmayer, D.B. and G.E. Likens. 2009. Monitoring for Ecological Knowledge. CSIRO Publishing (in press).

Lovett, G.M., D.A. Burns, C.T. Driscoll, J.C. Jenkins, M.J. Mitchell, L. Rustad, J.B. Shanley, G.E. Likens, and R. Haeuber. 2007. Who needs environmental monitoring? Frontiers in Ecology and the Environment 5(5):253–260.

Lovett, G.M., and M.J. Mitchell. 2004. Sugar maple and nitrogen cycling in the forests of eastern North America. Frontiers in Ecology and the Environment 2:81–88.

Lovett, G.M., K.C. Weathers, M.A. Arthur, and J.C. Schultz. 2004. Nitrogen cycling in a northern hardwood forest: Do species matter? Biogeochemistry 67:289–308.

Magnuson, J.J., D.M. Robertson, B.J. Benson, R.H. Wynne, D. Livingstone, T. Arai, R.A. Assel, R.G. Barry, V. Card, E. Kuusisto, N.G. Granin, T.D. Prowse, K.M. Stewart, and V.S. Vuglinski. 2000. Historical trends in lake and river ice cover in the Northern Hemisphere. Science 289:1743–1746.

McCabe, G.J., M.A. Palecki, and J.L. Betancourt. 2004. Pacific and Atlantic Ocean influences on multidecadal drought frequency in the United States. Proc. National Academy of Sciences 101(12):4136–4141.

McCabe, G.J., and D.M. Wolock. 2002. A step increase in streamflow in the conterminous United States. *Geophysical Research Letters* 29(24):2185, DOI:10.0129/ 2002GL015999.

McLauchlan, K.K., J.M. Craine, W.W. Oswald, P.R. Leavitt, and G.E. Likens. 2007. Changes in nitrogen cycling during the past century in a northern hardwood forest. Proc. of the National Academy of Sciences 104(18):7466–7470.

Meyer, J.L., and G.E. Likens. 1979. Transport and transformation of phosphorus in a forest stream ecosystem. Ecology 60(6):1255–1269.

Meyer, J.L., G.E. Likens, and J. Sloane. 1981. Phosphorus, nitrogen, and organic carbon flux in a headwater stream. *Arch. Hydrobiol.* 91(1):28–44.

Reynolds, R.C., Jr., and N.M. Johnson. 1972. Chemical weathering in the temperate glacial environment of the Northern Cascade Mountains. *Geochim. Cosmochim. Acta* 36:537–554.

Roskoski, J.P. 1980. Nitrogen fixation in a northern hardwood forest in the northeastern United States. *Plant and Soil* 54:33–44.

Saccone, L., D.J. Conley, G.E. Likens, S.W. Bailey, D.C. Buso, and C.E. Johnson. 2008. Factors that control the range and variability of amorphous silica in soils in the Hubbard Brook Experimental Forest. *Soil Sci. Soc. Am. J.* 72(6): 1637–1644.

Schindler, D.W. 1977. Evolution of phosphorus limitation in lakes. *Science* 195:260–262.

Siccama, T.G., T.J. Fahey, C.E. Johnson, T. Sherry, E.G. Denny, E.B. Girdler, G.E. Likens, and P. Schwarz. 2007. Population and biomass dynamics of trees in a northern hardwood forest at Hubbard Brook. *Can. J. For. Res.* 37:737–749.

Smith, V.H. 1998. Cultural eutrophication of inland, estuarine and coastal waters. In M.L. Pace and P.M. Groffman (eds.). *Success, Limitations and Frontiers in Ecosystem Science* (pp. 7–49). New York: Springer-Verlag.

Steinhart, G.S., G.E. Likens, and P.M. Groffman. 2000. Denitrification in stream sediments in five northeastern (USA) streams. *Verh. Internat. Verein. Limnol.* 27(3):1331–1336.

Templer, P.H., G.M. Lovett, K.C. Weathers, S.E. Findlay, and T.E. Dawson. 2005. Influence of tree species on forest nitrogen retention in the Catskill Mountains, New York, USA. *Ecosystems* 8:1–16.

Union of Concerned Scientists. 2006. Climate change in the U.S. Northeast. A Report of the Northeast Climate Impacts Assessment. www.northeast climateimpacts.org. 34 pp.

Ward, R.C., J.C. Loftis, and G.B. McBride. 1986. The "data-rich but information-poor" syndrome in water quality monitoring. *Environ. Management* 10(3):291–297.

Warby, R.A.F., C.E. Johnson, and C.T. Driscoll. 2009. Continuing acidification of organic soils across the northeastern USA: 1984–2001. *Soil Sci. Soc. Am. J.* 73:274–284.

Watmough, S.A., J. Aherne, C. Alewell., P. Arp, S. Bailey, T. Clair, P. Dillon, L. Duchnese, C. Eimers, I. Fernandez, N. Foster, T. Larssen, E. Miller, M. Mitchell, and S. Page. 2005. Sulfate, nitrogen and base cation budgets at 21 forested catchments in Canada, the United States, and Europe. *Environ. Monitor. Assess.* 109:1–36.

Yanai, R.D. 1990. The effect of disturbance by whole-tree harvest on phosphorus cycling in a northern hardwood forest ecosystem. Ph.D. Dissertation, Yale University. 194 pp.

7

SUMMARY AND CONCLUSIONS

THOMAS C. WINTER AND GENE E. LIKENS

Being a small lake in a mountainous setting, Mirror Lake has a highly dynamic interaction with the hydrologic cycle. The streams tributary to the lake drain steep mountainsides, so discharge into the lake responds rapidly to precipitation, especially rain, falling on the watershed. At the same time, when the elevation of the lake surface is near the elevation of the outlet dam, many episodes of large tributary inflow result in equally large surface outflow over the dam. The lake level declines below the top of the dam because of losses to evaporation and seepage to ground water. However, in most years, the decline is 30 cm or less.

The three tributaries contribute a long-term average of about 59 percent of the inflow to the lake, precipitation directly on the lake contributes about 26 percent, and ground water contributes the remaining 16 percent of the inflow. Even though surface outflow over the dam can be substantial at times of snowmelt and heavy rains, it is episodic and represents a long-term average of about 38 percent of the water lost from the lake. The largest loss of water from the lake is by way of seepage to ground water, which represents a long-term average of about 51 percent of the losses. The remaining 11 percent of the water lost is by evaporation.

In general, the largest surface inflows and outflows occur during the spring at times of snowmelt and spring rains. A secondary peak of surface inflows and outflows commonly occurs as a result of fall rains. Precipitation generally is higher during these seasons. At these times, surface outflow is the largest loss of water from the lake. However, because seepage to ground water is a fairly consistent and steady process year round, it is larger than surface outflow for most of the year, resulting in its being the largest long-term average loss of water from the lake. These dynamic exchanges of water result in the lake having an average residence time of only about a year.

The water-flow dynamics of Mirror Lake had a substantial effect on the nutrient dynamics of the lake. Solute transport at Mirror Lake using the simple mass balance approach revealed that the lake was generally a flow-through system for many watershed-derived solutes, where inputs approximately equaled outputs. However, it was also a sink for chemicals such as hydrogen ion, ammonium, nitrate, phosphate, and dissolved silica, which were largely retained in the system by internal limnological processes. This uptake, or neutralization, was proportionally very similar to the retention measured in long-term studies of terrestrial watersheds in the Hubbard Brook Valley.

Changes in the chemical inputs to Mirror Lake by way of precipitation and tributaries usually resulted in observable changes in the lake itself. Although the total nutrient inputs and outputs varied according to the quantities of water exchange, the actual mass balances were relatively insensitive to the hydrologic conditions observed over the 20 years. Wet years resulted in more total flux, but inputs and outputs were balanced. That the lake chemistry was a mixture of the three tributaries and groundwater chemistries was anticipated, but the degree to which direct precipitation contributed so dominantly for the most biologically reactive solutes, such as ammonium, nitrate, and phosphate, was not anticipated.

Estimates of groundwater inputs, based on a simple premise of uniform water transport characteristics and constant solute concentration, appears to have given reasonable closure to most of the budgets. In the one exception, where chloride appears to have been routinely released from the lake, we suspect that there are unmeasured areas of ground water having high chloride concentrations that could balance the chloride budget. An unmeasured cation (possibly reduced iron) would need to be involved to provide the necessary charge to balance the unsupported release of chloride from the lake, and this solution would need to enter the lake through a narrow point-source area undetected by the broader network of groundwater sample sites. The contribution of base cations and ANC in ground water was very effective in helping to neutralize acidic precipitation falling directly on the lake. Groundwater ANC provided 63 percent of the annual ANC influx, compared to 37 percent for all three tributaries. The largest individual solute flux was by way of seepage to ground water. For the purpose of estimating mass balances, it was critical to measure this largest mass of water and solutes.

The relative importance of ground water changed substantially with variations in climate. In dry years, when direct precipitation, stream inputs, and surface water outflow were smallest, the portion of solute mass flux and export by groundwater inflow and seepage outflow was maximized. Without an accurate estimate of flux for each flow path, even these simple observations would be highly speculative.

With respect to several physical and chemical characteristics important to the condition and well-being of Mirror Lake, the following observations were documented in this 20-year study.

Decreasing Sulfate Levels   The mass of sulfate in the lake is decreasing. In addition to decreasing sulfate loading from precipitation, the sulfate deposited infiltrates and is absorbed throughout the entire terrestrial and aquatic ecosystems. Groundwater influxes of ANC have made the lake less susceptible to the effects of atmospheric deposition by providing increased buffering in the lake.

Correlation of pH and Sulfate Levels with Precipitation   Increasing pH and decreasing sulfate concentration trends in the lake were correlated with similar changes in pH and sulfate fluxes from direct precipitation, and not with changes in any of the tributaries. Mirror Lake pH would have been between pH 7 and 7.5 prior to input and titration by anthropogenic atmospheric acids. Currently, the average pH of Mirror Lake is about 6.5, and so it has not fully "recovered" to its earlier state.

Sodium and Chloride Increases Due to Road Salting   Although the Northeast Tributary contributed only 3 percent of the annual average stream inflow to Mirror Lake, by 1994 it contributed about 50 percent of the average annual input of chloride. Equivalent ratios of sodium to chloride in the Northeast Tributary demonstrated the large and rapid impact of road salt applied to nearby Interstate Highway 93 on the lake. The ratio decreased quickly from about 3 to 0.8 when I-93 opened, reflecting sea salt sources, and it decreased further to about 0.4, reflecting the exchange of sodium ions for calcium and magnesium ions in the terrestrial watershed. Gradual declines in cation exchange capacity may be responsible for the sodium/chloride ratio rising back toward 0.8 since the mid-1980s. Concentrations of sodium and chloride in the lake did not decline in response to the decreases of salt in the Northeast Tributary after 1995; instead, total annual inputs of sodium and chloride to the lake continued to increase. The primary reason for this increase was increasing concentrations of sodium chloride from the Northwest and West Tributaries, which contributed 26 percent of the total water input to the lake. The annual use of about 720 Kg y^{-1} of sodium-chloride on local roads and driveways within the lake's watershed has a greater impact on Mirror Lake currently than the application of an average 28 metric tons of sodium-chloride per lane-mile on nearby I–93 since 1970.

Increasing Chemical Concentrations Data on chemical retention within Mirror Lake indicate that the lake retained phosphorus, nitrogen, dissolved silica, and hydrogen ion in all 20 years of this study. The variability in percent annual retention for most solutes was related to changes in tributary inputs, caused by retention or release in the terrestrial watershed. Retention of ammonium, nitrate, and phosphate in the watershed prevented serious changes in lake trophic conditions because it reduced the total influx to the lake to just a fraction of the total deposition to the entire watershed. Highs and lows in net imbalances of ammonium, hydrogen ion, nitrate, and phosphate were not related to wet or dry years, because these solutes were provided largely (>50%) by direct precipitation to the lake's surface.

The pattern in net release or retention for calcium, magnesium, sodium, potassium, and sulfate is primarily related to water movement through the lake. In wet years when water inputs exceed water losses, net retention values for these constituents generally approached zero or were negative (released).

The silica cycle was dominated by large tributary and groundwater inputs of dissolved silica (85 metric tons), which was balanced by substantial uptake and/or sedimentation (56 metric tons). About 66 percent of total, dissolved silica inputs were sequestered in the sediments during the period of this study.

Over 20 years, processes within the lake annually removed about 1.5 Kg of phosphate, 254 Kg of nitrate, 35 Kg of ammonium, 2044 Kg of dissolved silica, and 9030 equivalents of hydrogen ion from the lake's chemical mass.

During the 14,000 years of the lake's existence, it has filled approximately half full with sediment. For the first 13,800 years, the lake was relatively unproductive and nutrient poor, with water characterized by $Ca(HCO_3)_2$. Because of increased atmospheric loading from atmospheric acids, the chemistry of the lake gradually shifted to be dominated by $CaSO_4$ some 60 years ago, and currently, because of road salt input, the chemistry of the lake is characterized by calcium and sulfate-chloride. The lake remains unproductive and is expected to stay that way during the next 50 years or so because of the shoreline protection that has been done during the last 40 years or so. Nevertheless, accelerated changes in climate could substantially change this prediction.

It was possible to evaluate uncertainties in the various terms and overall amounts of both water and chemical budgets because of the completeness in measurement of all input and output terms and because of the long-term nature of these measurements. Such evaluation of uncertainties provides guidance to other scientists and managers regarding the cost and focus of effort to be devoted to lake studies.

INDEX

Note: Page references followed by a "t" or an "f" refer to tables or figures, respectively. For example, the reference 315f refers to a figure on page 315.

acid deposition, 70, 71, 84, 129–134. *See* acid rain
   Industrial Revolution, role of, 130
   lake changes due to, 131–134
   nitrates, 130–131
   pre-Industrial Revolution, 310
   recovery from effects of, 132–133
   and solute flux, 151
   sulfates, 130
acidity. *See also* pH
   increases in, 152
   temporal trends, 313
   water, 9
acid-neutralizing capacity (ANC), 10, 331
   analyses of, 72
   annual fluxes during driest/wettest years, 198t–199t
   annual volume-weighted concentration and year, analysis of relation between, 160t
   annual volume-weighted concentration, changes in, 317f
   annual volume-weighted concentrations vs. normal distribution, 159t
   budget, annual, 190t–191t
   chemical mass, 118
   groundwater inflow, 109–110, 111, 114, 135, 147, 152
   groundwater outflow, 77, 110, 111
   imbalances affecting, 133–134
   long-term volume-weighted concentrations, 246
   mass, changes in, 119, 196t
   measurement of, 155t
   Northeast Tributary, 76, 78, 87, 98–99, 109–111, 119, 135, 148, 153, 215, 239, 241, 254, 272t–273t, 304, 331–332
   percent contribution to total influx/outflow in driest/wettest year, 200t
   relation between annual volume-weighted concentration and year for atmospheric precipitation, tributaries, and surface outlet, 161t

acid-neutralizing capacity (ANC) (*continued*)
  relation between concentration and flow, 237
  seasonal fluxes, 125, 127
  solute budget, 253, 254
  surface water, 75
  in tributaries, 132
  20-year average annual inputs, 162t–163t
  20-year average annual outputs, 164t–165t
  20-year volume-weighted average/median concentrations and ion balances, 156t
  volume-weighted 20-year average or median concentrations for, 158t
acid precipitation. *See* acid rain
acid rain, 129, 130, 301, 304, 312–316, 318, 331. *See also* acid deposition
  documentation of, 316
  recovery from effects of, 134
agriculture, effects of, 304
air-land-water interactions, 305, 320
air temperature, average, 5
American toad (*Bufo americanus*), 9
ammonium, 331
  annual fluxes during driest/wettest years, 198t–199t, 332
  annual volume-weighted concentration and year, analysis of relation between, 160t
  annual volume-weighted concentrations vs. normal distribution, 159t
  average annual concentrations, 295t
  budget, annual, 174t–175t
  change in mass, 119
  concentration, changes in, 78, 244, 333
  groundwater inflow, 110
  groundwater outflow, 100, 111, 232
  input-output balance, 154
  long-term volume-weighted concentrations, 246
  mass, 101, 104f, 109, 119, 197t
  measurement of, 78, 155t, 229
  percent contribution to total influx/outflow in driest/wettest year, 200t
  in precipitation, 109, 110, 114, 121, 149
  relation between annual volume-weighted concentration and year for atmospheric precipitation, tributaries, and surface outlet, 161t
  relation between concentration and flow, 237
  retention, 119, 120, 140, 141, 143f, 149, 153, 332
  seasonal fluxes, 125, 149
  storage, change in, 250
  total inputs over 20 years, 140
  in tributaries, 98–99
  20-year average annual inputs, 162t–163t
  20-year average annual outputs, 164t–165t
  20-year volume-weighted average/median concentrations and ion balances, 156t
  uptake, 133
  volume-weighted 20-year average or median concentrations for, 158t
analytical Darcy method for calculating groundwater flux, 40, 211–214, 219
ANC. *See* acid-neutralizing capacity (ANC)
anemometer, 35
annual average total ionic charge, over time, 86f
annual solute imbalances, vs. annual water imbalances, 148f
annual volume-weighted concentrations for atmospheric precipitation, tributaries, and surface outlet of the lake, 161t

summary, 159t
and year, relation between using
  Kendall tau correlation coefficient
  analysis, 160t
and year, relation between using
  linear regression analysis, 160t
annual water imbalances, annual solute
  imbalances vs., 148
annual water volume, comparison of,
  296t
anoxia, 7
anthropogenic impact and change,
  318, 321
atmospheric deposition, 80, 120
  chemical characteristics of, 230, 306
  of nitrate, 130–131
  and solute flux, 151
  sulfate from, 130, 154
  of sulfur, 312 (*see also* acid rain)
  susceptibility to effects of, 331
  on top of lake ice, 234
atmospheric loading, 333
atmospheric precipitation, measurement
  of, 230
atmospheric radiation, 31
atmospheric water, 31, 33–37
  evaporation, 34–37
  precipitation, 33–34
average annual fluxes, budgets, and
  residuals, summary of, 194t–195t
average concentration time trends,
  annual volume-weighted, 88f–90f,
  92f, 95f–97f
"average" method for determining
  solute mass input from tributaries,
  237

balance. *See* input-output balance;
  nutrient concentrations and
  balances
base cation depletion, 318
bathymetric map, 2m
bedrock, 2, 24

outcrops, 27
piezometers, 41
topography, 26f
benthic algae, 15t
benthic bacteria, 15t, 18t
benthic fauna, 7
benthic fungi, 15t
benthic invertebrates, 7, 15t, 18t
bicarbonate, concentration of in
  tributaries, 91
biogeochemical characteristics, 299
biogeochemical flux, 11, 305–308, 320
biogeochemical processes
  changes in, over time, 69
  influencing the balance between
    acids and bases, 132
  wet and dry conditions, influence of,
    149–154
biological limnology, 7–10
biological productivity, 10, 144–145,
  312, 333
biotic transfer, 242
biotic uptake, 253
birds, 9, 15t
bladderwort (*Utricularia* sp.), 9
Books and Islands in Ojibwe Country, 322
Bowen-Ratio Energy Budget (BREB)
  method of measuring evaporation,
  34, 207, 221t
broad-crested weir, 209
brook trout (*Salvelinus fontinalis*), 9
brown bullhead (*Ictalurus nebulosus*), 9
brown trout (*Salmo trutta*), 9
bulk precipitation
  average ionic charge distribution for,
    85f
  chemical characteristics of, 84, 87
  trends, 94, 95f–97f
buried bedrock valleys, 24
burr weed (*Sparganium* sp.), 9

cadmium, 5
$Ca(HCO_3)_2$, 333

calcium, 333
  acid rain, effects of, 313
  annual fluxes during driest/wettest
    years, 198t–199t
  annual volume-weighted
    concentration and year, analysis of
    relation between, 160t
  annual volume-weighted
    concentration, changes in, 317f
  annual volume-weighted
    concentrations vs. normal
    distribution, 159t
  average annual concentrations,
    295t
  budget, annual, 166t–167t
  change in mass, 119
  concentrations, 10, 84, 85, 232
  decreasing concentrations of, 313,
    314f
  groundwater inflow, 110, 146, 242
  groundwater outflow, 99, 110, 111
  input-output balance, 114
  long-term volume-weighted
    concentrations, 246
  mass, 101, 102f, 118, 196t
  measurement of, 155t
  in Mirror Lake and Hubbard Brook
    Experimental Forest compared,
    314f
  percent contribution to total influx/
    outflow in driest/wettest
    year, 200t
  in precipitation, 109
  relation between annual volume-
    weighted concentration and year
    for atmospheric precipitation,
    tributaries, and surface outlet, 161t
  relation between concentration and
    flow, 237
  retention, 116, 142f, 144, 153, 242,
    332
  seasonal fluxes, 121, 125, 149
  solute budget for, 252, 255
  surface water outflow, 232
  in tributaries, 87, 91, 98, 114, 231
  20-year average annual inputs,
    162t–163t
  20-year average annual outputs,
    164t–165t
  20-year volume-weighted average/
    median concentrations and ion
    balances, 156t
  volume-weighted 20-year average or
    median concentrations for, 158t
calcium bicarbonate, 152
Campbell Scientific model CS500
  probe, 36
Campbell Scientific model 207 probe,
  36
carbon, average standing stock, 12, 18t
carbon budget, 214
carbon flux, 12f
$CaSO_4$, 333
catch efficiency, precipitation gauges, 206
catchment, 75, 306
cation, 313
  base, groundwater outflow, 77
  exchange, declines in, 153, 332
  unmeasured, 134, 154
Cayuga Lake, species diversity in, 9
chain pickerel (*Esox niger*), 9
charcoal, 5
chemical analyses, uncertainties in, 229
chemical budget(s), 112–113, 225–263.
  See also mass balance
  alternative approaches to
    determining, 236–246
  "average" method for determining
    solute mass input from tributaries,
    237
  budget patterns and processes, 256–257
  chemical analyses, uncertainties in, 229
  chemical mass input from
    groundwater, 241–243
  chemical mass input from tributaries,
    236–239

338 INDEX

chemical mass loss by flow through groundwater, 244, 246
chemical mass loss by surface outflow, 239–241
cumulative error in solute budgets, 252–256
to evaluate chemical retention or loss, 240
groundwater effects on, 71, 145–149
input-output balance, 112–114
"model" method for determining solute mass input from tributaries, 237
relation of inputs to outputs, 114
relation of uncertainties to hypotheses, 246–257
relations of water and solute budgets to lake concentrations, 257–262
residuals of inputs, outputs, and change in storage, 276t–277t
sample collection, uncertainties in, 230–236
solute budget residuals, 252
surface-water contributions, methods of determining, 278t–279t
uncertainties in, 262–263, 333
uses of, 240
water budgets used to determine, uncertainties in, 226–229
chemical characteristics
long-term trends in, 71
of water sources, 83–101
chemical composition of Mirror Lake water as percent, 316f
chemical concentrations
average annual, 295t
increasing, 332–333
chemical data, average, 17t
chemical flux, 69, 70
calculation of, 230
least-squares regression analysis, 236, 272t–273t

chemical input-output balance, watersheds, 119–120
chemical inputs
changes in, 330
daily, estimates of, 80–82
chemical limnology, 10–13
chemical mass
changes in, 112–113, 118–119
and chemical retention, 118–119
comparison, summary of, 274t–275t
input and output compared, 112
input-output imbalance and, 118, 247, 250
measurement of, uncertainties in, 118
chemical mass input, 236–239
annual solute mass in the tributaries and surface outlet of the lake, 238f
associated with hydrologic processes, 112
and chemical retention, comparison of on an annual basis, 282t–283t
concentrations used to represent groundwater chemical characteristics, 281t
from groundwater, 241–243
from tributaries, 236–239
chemical mass loss
calculating, 239
by flow through groundwater, 244, 246
by surface outflow, 239–241
chemical mass output, associated with hydrologic processes, 112
chemical output, daily, estimates of, 82
chemical release, 153
chemical retention, 115–117f, 153, 243f, 244f, 332–333
ammonium, 119, 120, 140, 141, 143f, 149, 153, 332
on an annual basis, comparison of, 280t, 282t–283t
annual trends by solute, 117f
calcium, 116, 142f, 144, 153, 242, 332
change in mass and, 118–119

INDEX 339

chemical retention *(continued)*
    chloride, 116, 142f, 144, 242
    defined, 115–116
    dissolved inorganic carbon, 144
    dissolved silica, 116, 140, 141, 146, 153, 332
    hydrogen ion, 116, 119, 120, 143f, 153, 332
    magnesium, 116, 142f, 144, 153, 242, 332
    nitrate, 120, 121–122, 131, 140, 141, 143f, 149, 153, 332
    nitrogen, 116, 119, 141, 154, 332
    phosphate, 119, 120, 140, 141, 149, 153, 242, 245f, 332
    phosphorous, 116, 141, 143f, 153, 332
    potassium, 116, 140, 142f, 153, 242, 245f, 332
    reasons for, 116
    sodium, 116, 142f, 144, 153, 242, 332
    sulfate, 116, 133, 142f, 144, 153, 242, 332
    in watersheds, 119–120
chemical "signature," 129
chemistry, 5, 10–13, 84
    human settlement, and changes in, 5
    long-term climate change and, 70, 120–129
chloride, 333
    annual fluxes during driest/wettest years, 198t–199t
    annual volume-weighted concentration and year, analysis of relation between, 160t
    annual volume-weighted concentration, changes in, 317f
    annual volume-weighted concentrations vs. normal distribution, 159t
    average annual concentrations, 295t
    average discrepancy (release), 134
    budget, annual, 182t–183t
    concentrations, 10, 84, 330
    groundwater inflow, 110, 111, 147–148
    groundwater outflow, 110, 116
    increases in, 152, 313
    input-output balance, 114
    isopleths of concentration with depth over time, 137f
    long-term volume-weighted concentrations, 246
    mass, 101, 106f, 118, 196t
    measurement of, 155t
    in Mirror Lake and Hubbard Brook Experimental Forest compared, 314f
    from Northeast Tributary, 148, 153, 332
    percent contribution to total influx/outflow in driest/wettest year, 200t
    relation between annual volume-weighted concentration and year for atmospheric precipitation, tributaries, and surface outlet, 161t
    relation between concentration and flow, 237
    retention, 116, 142f, 144, 242
    and road salt contamination, 10, 70, 119–120, 134–140, 313, 317, 321, 331–332, 333
    seasonal fluxes, 121, 125, 149
    solute budget, 253, 255
    solute imbalance vs. water imbalance, 148f
    storage, variability in, 119
    from tributaries, 87, 91, 98, 109, 114, 148, 153, 231
    20-year average annual inputs, 162t–163t
    20-year average annual outputs, 164t–165t
    20-year volume-weighted average/median concentrations and ion balances, 156t
    volume-weighted 20-year average or median concentrations for, 158t

Clean Air Act, emission controls, 131
clear cutting, 69
climate, 2, 4
   accelerated changes in, 333
   effects of, 152–153
   influences of on chemical seasonal patterns, 120–129
   water budget, influences of, 46–50
climate change, 301, 318, 320, 333
   long-term vs. short-term, 70
   warming, 302
common loon (*Gavia immer*), 9
common snappers (*Chelydra serpentine*), 9
concentrations. *See* nutrient concentrations and balances
conductivity, 10
coniferous vegetation, 5
cultural development, 320
cultural eutrophication, 301, 312–316
   acid rain, 129, 130, 134, 304, 312–316, 318

dam, 2, 31, 70, 209
deciduous vegetation, 5
deicers, contamination by, 70
depth-time temperature isopleths for dry/wet years, 8f
development, protection against, 4
DIC. *See* dissolved inorganic carbon (DIC)
dissolved inorganic carbon (DIC)
   analyses of, 72
   annual fluxes during driest/wettest years, 198t–199t
   annual volume-weighted concentration and year, analysis of relation between, 160t
   annual volume-weighted concentrations vs. normal distribution, 159t
   budget, annual, 188t–189t
   groundwater inflow, 109–110, 111, 147
   groundwater outflow, 110, 111
   increased loading from soil disturbance and agriculture, 304
   long-term volume-weighted concentrations, 246
   mass, changes over time, 197t
   measurement of, 155t
   Northeast Tributary, 76
   percent contribution to total influx/outflow in driest/wettest year, 200t
   relation between annual volume-weighted concentration and year for atmospheric precipitation, tributaries and surface outlet, 161t
   retention, 144
   solute budget, 253, 255
   storage, variability in, 119
   surface water, 75
   20-year average annual inputs, 162t–163t
   20-year average annual outputs, 164t–165t
   volume-weighted 20-year average or median concentrations for, 158t
dissolved inorganic nitrogen, 308–311
   concentration and inputs of, 310f
   flux of, 310
   human activity and, 312
   sources of, 308–309
dissolved organic carbon (DOC), 16t
   analyses of, 72
   annual fluxes during driest/wettest years, 198t–199t
   average concentration, 10
   budget, annual, 186t–187t
   groundwater inflow, 110, 114
   groundwater outflow, 110
   mass, changes over time, 197t
   measurement of, 155t, 229
   percent contribution to total influx/outflow in driest/wettest year, 200t
   relation between annual volume-weighted concentration and year for atmospheric precipitation, tributaries, and surface outlet, 161t

dissolved organic carbon (DOC)
(*continued*)
  relation between concentration and
    flow, 237
  solute budget, 252, 255
  in tributaries, 114
  20-year average annual inputs,
    162t–163t
  20-year average annual outputs,
    164t–165t
  volume-weighted 20-year average or
    median concentrations for, 158t
dissolved silica, 331
  annual volume-weighted
    concentration and year, analysis of
    relation between, 160t
  annual volume-weighted
    concentrations vs. normal
    distribution, 159t
  average annual concentrations, 295t
  budget, annual, 192t–193t
  change in mass, 119
  concentration, changes in, 333
  concentrations, simulation of, 239,
    240
  flux, 311–312
  groundwater inflow, 114, 147
  groundwater outflow, 111, 114
  long-term volume-weighted
    concentrations, 246
  mass, 119
  measurement of, 155t
  relation between annual volume-
    weighted concentration and year
    for atmospheric precipitation,
    tributaries, and surface outlet, 161t
  relation between concentration and
    flow, 237
  retention, 116, 140, 141, 146, 153,
    332
  seasonal fluxes, 121, 122, 125, 149
  storage, change in, 250
  total inputs over 20 years, 140
  from tributaries, 231
  20-year average annual inputs,
    162t–163t
  20-year average annual outputs,
    164t–165t
  volume-weighted 20-year average or
    median concentrations for, 158t
dissolved solids, estimated concentration
  of, 305f
dissolved substance flux, 69
DOC. *See* dissolved organic carbon
  (DOC)
drainage, 4
drainage basin, 1, 5
drought conditions, 307
dry-bulb thermistor, 35

An Ecosystem Approach to Aquatic Ecology:
  Mirror Lake and Its Environment, 13
ecosystems, difficulties in study of, 13
ecosystem services, diminishment of,
  319
eddy correlation method of measuring
  evaporation, 207
El Niño Southern Oscillation (ENSO)
  and wet years, relation between,
  307–308
energy budget, ice cover and, 5
environmental impacts, 301
environmental problems, identification
  of, 319, 320
Erdich, Louise, 322
erosion, 317, 321
eutrophication, 301, 312–316
evaporation, 28, 31, 45
  Bowen-Ration Energy Budget
    (BREB) method of measuring, 34,
    207, 221t
  eddy correlation method of
    measuring, 207
  long-term monitoring of, 207
  mass transfer method of measuring,
    207–208

mean monthly, 51f, 56
measurement of, 34–37, 207–208, 221t
Priestly-Taylor method of measuring, 34, 36, 208
rates, averaged, 37
ratio of total outflow to at hydrologic steady state, 215–216
20-year average annual outputs, 164t–165t
uncertainties in, 207–208
evapotranspiration
changes in, 321
Hubbard Brook Experimental Forest, 12, 28, 69

fall rains, 329
filtration method of water sample collection, 73
fire, 5
first-order error analysis, 43–44, 227
fish, 9, 15t, 18t
stocking, 9
flooding, 321
flushing time, 7
flux, 14
annual, during driest/wettest years, 198t–199t
average annual, summary of, 194t–195t
food web, 12f, 13
forest removal, 70
freezing rain, 321
functional integrity, 321

gaseous exchange, 242
geologic deposits, 211
glacial advances, 24, 302
glacial deposits, 2, 24
heterogeneity associated with, 212
piezometers in, 41
thickness of, 25f, 212
glacial retreat, 5, 301, 311

glacial till, 24f, 28, 76, 212
global warming, 5, 7, 302
green frog (*Rana clamitans*), 9
groundwater, 23, 28, 31, 40–43, 211–220
acid-neutralizing capacity (ANC), 147, 152, 331
annual contribution to Mirror Lake, 262
basin simulation model, discretization grid for, 218f
chemical characteristics, concentrations used to represent, 281t
chemical mass input from, 241–243
chemical mass loss by, 244, 246
and climate variations, 331
discharge, 29, 215, 219
effects of on chemical budgets, 71, 145–149
flux, measurement of, 211–216, 219
"hotspots," 149
lake flow into, 234–235
numerical model of inflow from groundwater basin, 217–220
pH, 77
sample collection, 235–236
seepage of lake water to, 29, 40, 57, 82, 330
seepage of, to lake, 40, 77, 154
system, thickness of, 213
values, error in, 220
volume of discharge, 215
groundwater flow
equation, 217
Hubbard Brook Experimental Forest, 69
from lake, 234–235
into lake, 235–236
groundwater flux, 149
calculation, segments of shoreline associated with, 213f

groundwater flux *(continued)*
    contribution to lakes, estimation of, 261
    methods for calculating, 40, 211–216, 219
    uncertainty in, 229, 255
groundwater inflow (GWI), 50, 109–110, 229, 329, 331
    acid-neutralizing capacity (ANC), 109–110, 111, 114, 147, 152
    alternative methods to determine, 219
    ammonium, 110
    analytical Darcy method for determining, 40, 211–214, 219
    average, 44, 150–151, 229
    calcium, 110, 146, 242
    chemistry, 91
    chloride, 110, 111, 147–148
    dissolved inorganic carbon (DIC), 109–110, 111, 147
    dissolved organic carbon (DOC), 110, 114
    dissolved silica, 114, 147
    effects of on chemical budgets, 146–147
    estimates of, 81, 330
    hydrogen ion, 110, 242
    from lake, 234–235
    into lake, 235–236
    magnesium, 110, 146, 242
    nitrate, 110
    numerical model of, 217–220
    nutrient concentrations, trends in, 99
    oxygen isotope method for determining, 40, 214–216, 219, 222t
    phosphate, 110
    potassium, 110, 114, 147
    sample collection, 76–77
    seasonal patterns of, 150–151, 154
    and seepage outflow compared, 29
    sodium, 110, 111, 147–148, 242
    and solute imbalances, 114
    sulfate, 110, 114, 130, 242
groundwater outflow (GWO), 57, 64, 331
    acid-neutralizing capacity (ANC), 77, 110, 111
    ammonium, 100, 111, 232
    average, 44
    calcium, 99, 110, 111
    chemistry, 91, 93
    chloride, 110, 116
    dissolved inorganic carbon (DIC), 110, 111
    dissolved organic carbon (DOC), 110
    dissolved silica, 111, 114
    effects of on chemical budgets, 145–146
    flux, 150
    hydrogen ion, 110, 111
    magnesium, 99, 110, 111
    nitrate, outflow, 77, 100, 110, 111
    nutrient concentrations, trends in, 99–101
    phosphate, 100, 110, 111
    potassium, 110, 111
    ratio of to evaporation, 215
    sample collection, 77
    seasonal patterns of, 57, 154
    sodium, 99, 110
    sulfate, 99–100, 110, 111
    trends, 99–101
grout basket, 41

habitat loss, 318
hemlock trees (*Tsuga canadensis*), 303
hickory (*Carya* spp.), 318
highway
    construction of, 134–135
    ecological effects of, 138, 140, 153, 331–332, 333
    improvements to prevent salt infiltration, 137–138, 321
    property losses due to, 139

road salt contamination, 10, 70,
119–120, 134–140, 153, 315,
317–318, 321
runoff associated with, 135, 140,
315
salt application records, 138
historical change, 302–305
housing units, 4
Hubbard Brook, 2, 45
Hubbard Brook Ecosystem Study, 4, 11,
46, 72, 230
Hubbard Brook Experimental Forest,
12, 72, 85, 130, 230, 313
Hubbard Brook Experimental Forest
Watershed, 46, 69
Hubbard Brook Valley, 2, 24, 55, 330
Hubbard Brook Web site, 72
human activity, 301
effects of, 70, 152–153, 321
modification of biogeochemical
fluxes by, 312
and presence of toxic metals, 5
human settlement, 5
hurricanes, effects of
Dennis, 55
Floyd, 55
hydraulic conductivity, 43, 77, 211–212,
214, 219, 220, 229
hydraulic gradient, determination of,
229
hydraulic head, 209
hydrogen ion, 331
annual fluxes during driest/wettest
years, 198t–199t, 332
annual volume-weighted
concentration and year, analysis of
relation between, 160t
annual volume-weighted
concentration, changes in, 317f
annual volume-weighted
concentrations vs. normal
distribution, 159t
average annual concentrations, 295t

budget, annual, 176t–177t
change in mass, 119
concentration, changes in, 333
groundwater inflow, 110, 242
groundwater outflow, 110, 111
imbalances, 133
input-output balance, 154
mass, 101, 104f, 119, 196t
percent contribution to total influx/
outflow in driest/wettest year,
200t
in precipitation, 110, 114, 120, 121,
131
relation between annual volume-
weighted concentration and year
for atmospheric precipitation,
tributaries, and surface outlet, 161t
relation between concentration and
flow, 237
retention, 116, 119, 120, 143f, 153,
332
seasonal fluxes, 149
storage, change in, 250
20-year average annual inputs,
162t–163t
20-year average annual outputs,
164t–165t
20-year volume-weighted average/
median concentrations and ion
balances, 156t
volume-weighted 20-year average or
median concentrations for, 158t
hydrogeologic cross-section, 27f
hydrogeologic setting, 24–27
hydrological and biogeochemical fluxes,
301, 305–308, 320
hydrologic cycle, interactions with,
329
hydrologic flux, 305–308, 320
into and out of Mirror Lake, 11, 31,
306f
balance between solute input and loss
by, 256–257

hydrologic "memory," 52
hydrologic pathways, and salt
    contamination, 71, 139–140
hydrologic processes, 27–31
    chemical mass input and output
        associated with, 112
hydrologic steady state, 220
    ratio of total outflow to evaporation
        at, 215–216
    water budget equation associated
        with, 214–215
hydrologic variables, locations of
    instruments to measure, 32f
hydrology, 13
hypotheses, 225
    relation of uncertainties to, 246–257

ice block, 4
ice cover
    duration of, 5, 6f, 320
    effect of on annual lake storage
        estimates, 233–234
    effect of on nutrient inputs, 80
    melting, 5, 6f, 7, 27
"ice houses," 11f
ice melt, 4, 6f, 80
    dilution from, 17t
index wells, 212, 235
influx/outflow during driest/wettest
    years, percent contribution to total,
    200t
injection and withdrawal test, 43
inorganic matter, sedimentation rates,
    309f
input-output balance
    acidic deposition and, 133
    ammonium, 154
    calcium, 114
    calculating, 112–114
    chemical mass represented by, 247
    and chemical retention, 116, 243
    chloride, 114
    defined, 112

hydrogen ion, 154
    and lake storage changes, time trends
        in, 102f–108f
    magnesium, 114, 146
    nitrate, 154
    phosphate, 154
    positive vs. negative, 112
    "pulse" effect, 121
    in relations to change in mass in
        storage, 247, 249
    sodium, 114, 146
    in watersheds, 119–120
inputs, total, 140
Interstate 93 (I-93), 3f, 4, 139, 319,
    321
    ecological impact of, 4, 134
invasive species, 301, 318
ionic charge
    average distribution for bulk
        precipitation, 85f
    change, average total over time, 86f
iron, 5, 134, 154
Isoetes sp., 9

Kendall tau test, use of, 83, 93, 94, 110,
    160t

Lady Slipper Road, 139
lakebed sediment, distribution of, 10f
lake ecosystem function, changes in,
    319
lake integrity, need to protect, 322
lake mass
    calculation of, 79
    chemical changes, 79–80
lake stage
    daily average, 54f
    daily to monthly compared, 63f
    decline, 47, 50
    mean monthly, 51f, 66t
lake surface, elevation, 329
lake volume
    annual, comparison of, 296t

changes in, 31–33, 57–58, 118
rainfall, and cumulative change in, 53f
represented by water balance and cumulative error in measured water balance, 265t
lake water, 233–235
  groundwater flow into, 235–236
  ice effects on annual storage estimates, 233–234
  lake flow into groundwater, 234–235
  losses of, via surface water and groundwater, 57
  samples, collection of, 73–74, 233–235
  temperature and oxygen profiles, 73
lead, 5
least-squares regression model, use of, 236, 272t–273t
limnology
  biology, 7–10
  chemical, 10–13
  history, 4–5
linear regression models, use of, 82, 160t

macrophytes, 9, 10, 15t, 18t, 31
macropores, 28
magnesium
  acid rain, effects of, 313
  annual fluxes during driest/wettest years, 198t–199t
  annual volume-weighted concentration and year, analysis of relation between, 160t
  annual volume-weighted concentration, changes in, 317f
  annual volume-weighted concentrations vs. normal distribution, 159t
  average annual concentrations, 295t
  budget, annual, 168t–169t, 229
  change in mass, 119
  concentrations, 85, 232
  decreasing concentrations of, 313, 314f
  groundwater inflow, 110, 146, 242
  groundwater outflow, 99, 110, 111
  input-output balance, 114, 146
  long-term volume-weighted concentrations, 246
  mass, 101, 102f, 118, 196t
  measurement of, 155t
  in Mirror Lake and Hubbard Brook Experimental Forest compared, 314f
  monthly lake mass and input/loss compared, 128f
  percent contribution to total influx/outflow in driest/wettest year, 200t
  in precipitation, 109
  relation between annual volume-weighted concentration and year for atmospheric precipitation, tributaries, and surface outlet, 161t
  relation between concentration and flow, 237
  residual and cumulative error of budget compared, 293t
  retention, 116, 142f, 144, 153, 242, 332
  seasonal fluxes, 121, 125, 149
  solute budget, 252, 255
  solute imbalance vs. water imbalance, 148f
  surface water outflow, 232
  in tributaries, 91, 114
  20-year average annual inputs, 162t–163t
  20-year average annual outputs, 164t–165t
  20-year volume-weighted average/median concentrations and ion balances, 156t
  volume-weighted 20-year average or median concentrations for, 158t
mallard ducks, 9
mammals, 15t
manganese, 5, 134

mass balance, 71, 112, 225
  budget, 113
  for ecosystem, description of, 11–12
  estimating, 145
  uncertainty and, 225
mass in the lake, differences in over time, 196t–197t
mass transfer method of measuring evaporation, 207–208
Mirror Lake
  as an aquatic ecosystem, 13
  chemical characteristics of, 70
  chemistry and productivity of, 5
  cultural history, 1
  formation of, 4, 302
  future, prospects for, 316–322
  historical change and, 302–305
  limnological history of, 4–5, 13
  location, 1
  names, 1
  scientific study, reasons for, 13, 23, 301, 308–312
  similarity of to other small- to medium-sized lakes, 301
  size and depth, 1
  today, overview of, 5–13
  trends, 93–94
Mirror Lake Hamlet, 4
"model" method for determining solute mass input from tributaries, 237
morphometric characteristics, 15t

National Weather Service, 46
natural ecosystem biogeochemical processes, changes in over time, 69
net ecosystem budget, 12
net ecosystem flux, 225
net gaseous flux, 225
neutralization, 330
New Hampshire climate division, 46
nitrate, 331
  annual fluxes during driest/wettest years, 198t–199t, 332
  annual volume-weighted concentration and year, analysis of relation between, 160t
  annual volume-weighted concentrations vs. normal distribution, 159t
  assimilatory reduction, 133
  atmospheric deposition of, 130–131
  average annual concentrations, 295t
  average concentration of, 131
  budget, annual, 180t–181t
  change in mass, 119
  concentration, changes in, 333
  concentrations, monthly volume-weighted average, 122f
  groundwater inflow, 110
  groundwater outflow, 77, 100, 110, 111
  input-output balance, 154
  mass, 105f, 109, 119, 197t
  mass, monthly, 123f
  mass input and mass loss, monthly differences between, 123f
  measurement of, 155t
  in Mirror Lake and Hubbard Brook Experimental Forest compared, 314f
  percent contribution to total influx/outflow in driest/wettest year, 200t
  in precipitation, 109, 110, 114, 121, 131, 149
  relation between annual volume-weighted concentration and year for atmospheric precipitation, tributaries, and surface outlet, 161t
  relation between daily flow and concentration, 237
  retention, 120, 121–122, 131, 140, 141, 143f, 149, 153, 332
  seasonal fluxes, 121–122, 125, 149
  storage, change in, 250
  total inputs over 20 years, 140
  in tributaries, 98, 99

20-year average annual inputs,
162t–163t
20-year average annual outputs,
164t–165t
20-year volume-weighted average/
median concentrations and ion
balances, 156t
volume-weighted 20-year average or
median concentrations for, 158t
nitrogen
average standing stock, 12, 18t
concentrations, 10
deposition, 309
retention, 116, 119, 141, 154, 332
sedimentation rates, 309f
sediment deposition of, 309
nitrogen cycline, 309
nitrous oxide (NO_x), emissions of, 130–131, 134, 313
Northeast Tributary, 2
ammonium in, 78
diversion of for highway construction, 4, 135
inflow/outflow, 76
nutrient concentrations, trends in, 98
road salt contamination of, 10, 111, 119, 138
sodium and chloride concentrations, 135–136, 138–139, 148, 153, 332
water budget, contribution to, 28
water chemistry, 87
watershed, 304
Northwest Tributary, 2, 28, 87, 98, 139, 153, 226, 304, 332
numerical simulation method for calculating groundwater flux, 40
nutrient budget model to determine average solute concentration, 258
nutrient concentrations and balances, 9, 301. *See also individual nutrients by name*
average annual for 1981–2000, 295t
of calcium, magnesium, potassium, sulfate, nitrate, and pH compared, 314f
changes in and their causes, 71, 102f–108f, 140–145
essential or limiting nutrients, predictions for retention of, 141, 144
non-essential or non-limiting nutrients, predictions for retention of, 144
relation to fluxes in water and solutes, 110–111
total inputs, 140, 141
trends in, 88f–90f, 92f, 93–94, 101
volume-weighted average (VWA), 80, 81, 88f–90f, 92f
nutrient content, 9
nutrient dynamics, 69–200, 330
acidic deposition, response, and recovery, 70, 71, 129–134
analytical problems in study of, 78–79
chemical budgets, groundwater effects on, 71, 145–149
chemical characteristics of water sources, 71, 83–101
climate influences on chemical seasonal patterns, 120–129
data calculation and assumptions, 79–83
hydrologic pathways and salt contamination, 134–140
nutrient concentrations and balances, changes in, 71, 93–94, 140–145
research methods, 71–83
sample collection, 72–77
solute mass and fluxes, 101–120
wet and dry conditions, biogeochemical influence of, 71, 149–154
nutrient inputs, 70, 80, 330

oak (*Quercus* spp.), 318
oligotrophic, 4, 144, 312
organic carbon flux, 12f
organic carbon inputs, 10, 16t
organic matter, sedimentation rates, 309f
outlet streamwater, 91, 93
outlet surface water, trends in, 99–101
output, daily, estimates of, 82
overturn, 5, 7

painted turtles (*Chryscarys picta*), 9
Palmer Drought Severity Index (PDSI), 46
Palmer Hydrologic Drought Index (PHDI), 46–47, 48f, 49f
parameter-estimated hydraulic conductivity, 219
Parshall flume, 37, 38f, 39, 64, 208, 210
particulate matter input/output, 242, 253
Pearson's *r* test, use of, 83, 111
pelagic algae (phytoplankton), 7, 15t
pelagic bacteria, 15t
pelagic fungi, 15t
pelagic zooplankton, 15t
Pemigewasset River, 27, 307
peristaltic pump, use of, 73, 77
permanent residences, 4
pH, 5, 10, 72
   average, in precipitation, 84, 131, 132
   groundwater, 77
   lake average, 152
   measurement of, 155t
   in Mirror Lake and Hubbard Brook Experimental Forest compared, 314f
   Northeast Tributary, 76
   and precipitation, correlation with, 331
   surface water, 75
   20-year volume-weighted average/median concentrations and ion balances, 156t
   volume-weighted 20-year average or median concentrations for, 158t
phosphate, 331
   annual fluxes during driest/wettest years, 198t–199t, 332
   annual volume-weighted concentration and year, analysis of relation between, 160t
   annual volume-weighted concentrations vs. normal distribution, 159t
   average annual concentrations, 295t
   budget, annual, 184t–185t
   concentration, changes in, 79, 244, 333
   groundwater inflow, 110
   groundwater outflow, 100, 110, 111
   input-output balance, 154
   long-term volume-weighted concentrations, 246
   mass, 101, 106f, 109, 119, 197t
   measurement of, 78–79, 155t
   percent contribution to total influx/outflow in driest/wettest year, 200t
   in precipitation, 109, 110, 114, 121, 149
   precipitation, tributaries, and surface outlet, 161t
   relation between annual volume-weighted concentration and year for atmospheric retention, 119, 120, 140, 141, 149, 153, 242, 245f, 332
   seasonal fluxes, 125, 149
   storage, variability in, 119, 250
   total inputs over 20 years, 140
   in tributaries, 98, 99
   20-year average annual inputs, 162t–163t
   20-year average annual outputs, 164t–165t
   20-year volume-weighted average/median concentrations and ion balances, 156t

volume-weighted 20-year average or
  median concentrations for, 158t
phosphorus
  average standing stock, 12, 18t
  budget, 13
  concentrations, 10, 311
  deposition, 309
  flux, 311
  human activity and, 312
  relation between concentration and
    flow, 237
  retention, 116, 141, 143f, 153, 332
  sedimentation rates, 309f
physical limnology, 5–7
physical setting, 23
physiographic setting, 23
phytoplankton, 7, 10, 15t, 18t
piezometer, 40–41
  bedrock, 41
  construction, 42f
  in glacial deposits, 41
  nest, 41
pipewort (*Eriocaulon* sp.), 9
Pleasant View Farm site, 34
pondweed (*Potamogeton* sp.), 9
potassium
  acid rain, effects of, 313
  annual fluxes during driest/wettest
    years, 198t–199t
  annual volume-weighted
    concentration and year, analysis of
    relation between, 160t
  annual volume-weighted
    concentration, changes in, 317f
  annual volume-weighted
    concentrations vs. normal
    distribution, 159t
  average annual concentrations, 295t
  budget, annual, 170t–171t
  correlation of with pH and sulfate
    levels, 331
  decreasing concentrations of, 313,
    314f

groundwater inflow, 110, 114, 147
groundwater outflow, 110, 111
long-term volume-weighted
  concentrations, 246
mass, 101, 103f, 118, 196t
measurement of, 155t
in Mirror Lake and Hubbard Brook
  Experimental Forest compared,
  314f
percent contribution to total influx/
  outflow in driest/wettest year, 200t
relation between annual volume-
  weighted concentration and year
  for atmospheric precipitation,
  tributaries, and surface outlet, 161t
relation between concentration and
  flow, 237
relation between inputs and outputs,
  243
retention, 116, 140, 142f, 153, 242,
  245f, 332
seasonal fluxes, 121, 125, 149
solute budget, 252, 255
storage, variability in, 119
total inputs over 20 years, 140
in tributaries, 91
20-year average annual inputs,
  162t–163t
20-year average annual outputs,
  164t–165t
20-year volume-weighted average/
  median concentrations and ion
  balances, 156t
volume-weighted 20-year average or
  median concentrations for, 158t
precipitation, 28, 33–34, 69, 312,
  329–330
  acidic (*see* acid rain)
  annual, 2, 4, 46
  atmospheric, measurement of on an
    event basis, 230
  average pH in, 84, 131, 132, 152
  bulk, characteristics of, 84, 87

precipitation *(continued)*
  bulk collectors, 75, 230
  concentrations correlated with elevation, 230–231
  correlation of pH and sulfate levels with, 331
  cumulative, 53f
  daily, estimates of, 80–82
  decreases in, 321
  depth-time temperature isopleths for dry/wet years, 8f
  direct, effect of on chemical budgets, 154
  distribution, 55
  fluxes, daily, 81
  gauges for measuring, 34f, 75, 206–207
  Hubbard Brook Experimental Forest, 69, 75
  hurricanes, 55
  influxes, 109
  inter-annual variability in, 50–55
  mean monthly, 51f, 56
  measurement of, 33–34, 206
  month-to-month variability in, 58–64
  Palmer Hydrologic Drought Index (PHDI), 46–47, 48f, 49f
  sample collection, 75, 230–231
  seasonal variability in, 55–58, 120–121
  and solute balance, contributions to, 114
  solutes, changes in, 316
  trends, 94, 152
  uncertainties in, 206–207
  wet vs. dry years, 151
  winter, changes in, 321
"pressure points," 320
Priestly-Taylor equation, 37
Priestly-Taylor method of measuring evaporation, 34, 36, 208
productivity, 5, 10, 144–145, 310, 333

protected shoreline, 4
pyranometer, 36
pyrgeometer, 36

quality control, research results, 83

radiation sensors, 38f
raft station, 35f
rainbow trout (*Oncorhynchus mykiss*), 9
rainfall. *See* precipitation
recreational activities, 4
recreational fishery, 9
red-spotted newt (*Notophthalmus v. viridescens*), 9
regression analysis, use of, 82–83, 94
release, chemical, 332
research methods
  ammonium measurements, 78
  data calculations and assumptions, 79–83
  input/output estimates, 80–82
  phosphate measurements, 78–79
  and precision, 155t
  quality-control/quality-assurance protocols, 71–72
  routine sampling program, 72
  sample collection, 72–77
  specific analytical problems, 78–79
  statistical analyses, 82–83
research results, 83–200
  acidic deposition, response, and recovery, 129–134
  annual input-output balances and lake storage changes, time trends, 102f–108f
  biogeochemical influence of wet and dry conditions, 149–154
  changes in nutrient concentrations and balances and their causes, 140–145
  chemical characteristics of water sources, 83–101
  climate influences on chemical seasonal patterns, 120–129

groundwater effects on chemical budgets, 145–149
hydrologic pathways and salt contamination, 134–140
questions raised by, 129
solute mass and fluxes, 101–120
tables summarizing, 155–200
residuals, average annual, summary of, 194t–195t
retention
   annual trends by solute, 117f
   change in mass and, 118–119
   defined, 115–116
   of essential or limiting nutrients, predictions for, 141, 144
   of non-essential or non-limiting nutrients, predictions for, 144
   patterns in over time, 332
   solute, annual trends in, 117f
road salt
   anti-caking ingredient, 139
   application records, 138
   contamination, 10, 70, 119–120, 134–140, 315, 317–318, 319, 321
   effects of, 138, 153, 331–332, 333
   infiltration, efforts to prevent, 137–138
Robert S. Pierce Ecosystem Laboratory, 34, 38f, 72, 75, 77, 136

salinization, 318
salt. *See* road salt
sample collection, 72–77
   filtration methods, 73
   incoming ground water, 76–77, 235–236
   lake water, 73–74, 233–235
   outflowing ground water, 77
   peristaltic pump, use of, 73, 77
   precipitation, 75, 230–231
   surface outflow, 232–233
   surface water, 75–76, 231–235
   temperature and oxygen profiles, 73

tributary inflow/outflow, 75, 231–232
uncertainties in, 230–236
wells used to sample ground water, locations of, 74f
sediment, 2, 4–5, 18t, 27, 154, 333
   chemistry of, 70
   lakebed, distribution of, 10f
   minerals contained in, 5
   nitrogen deposition, 309
sedimentation, 12f, 332
   estimated historical rates of, 303f
   rates, 309f
seepage outflow, 29, 31, 77
Shapiro-Wilk test, use of, 83, 93, 240
shoreline
   degradation, 319
   development of, 318, 319
   protection, 4, 320, 322, 333
SigmaStat® 3.5 software, 82
silica, concentrations of, 239, 311–312. *See also* dissolved silica
silica cycle, 154, 332–333
silicate
   annual fluxes during driest/wettest years, 198t–199t
   concentrations, monthly volume-weighted average, 124f
   mass, changes over time, 197t
   mass input and mass loss, monthly differences between, 124f, 125f, 128f
   percent contribution to total influx/outflow in driest/wettest year, 200t
   seasonal fluxes, 121
   in tributaries, 114
   uptake, 144
single-well hydraulic "slug" tests, 43, 212, 214, 220
sleet, 319
smallmouth bass (*Micropterus dolomieui*), 9
snowmelt, 329

INDEX 353

sodium
  annual fluxes during driest/wettest years, 198t–199t
  annual volume-weighted concentration and year, analysis of relation between, 160t
  annual volume-weighted concentration, changes in, 317f
  annual volume-weighted concentrations vs. normal distribution, 159t
  average annual concentrations, 295t
  budget, annual, 172t–173t
  change in mass, 119
  concentrations, 10, 84, 85, 140
  groundwater inflow, 110, 111, 147–148, 242
  groundwater outflow, 99, 110
  increases in, 152, 313
  input-output balance, 114, 146
  long-term volume-weighted concentrations, 246
  mass, 101, 103f, 118, 196t
  measurement of, 155t
  in Mirror Lake and Hubbard Brook Experimental Forest compared, 314f
  percent contribution to total influx/outflow in driest/wettest year, 200t
  in precipitation, 109
  relation between annual volume-weighted concentration and year for atmospheric precipitation, tributaries, and surface outlet, 161t
  retention, 116, 142f, 144, 153, 242, 332
  and road salt contamination, 10, 70, 119–120, 134–140, 313, 317, 321, 331–332, 333
  seasonal fluxes, 121, 149
  solute budget, 252, 255
  solute imbalance vs. water imbalance, 148f
  from tributaries, 87, 91, 94, 109, 114, 153, 231
  20-year average annual inputs, 162t–163t
  20-year average annual outputs, 164t–165t
  20-year volume-weighted average/median concentrations and ion balances, 156t
  volume-weighted 20-year average or median concentrations for, 158t
soil disturbance, effects of, 304
solar radiation, 31
solute(s)
  concentrations, trends in, 93
  input, trends for all sources of, 145f
  retention/release, short-term climate change and, 70
  seasonal changes and variations in, 125, 127, 129
  selected, monthly volume-weighted average concentrations of, 126f–127f
  storage, wet vs. dry years, 151
  watershed, retention trends, 142–143
solute balance, 248f–250f, 259f–260f
solute budgets
  balance between solute input and loss by hydrologic fluxes, 256
  components, contribution of to cumulative error, 290t–292t
  cumulative error in, 252–256
  expression of, 113
  groundwater contribution to, 113
  patterns and processes, 256–257
  relation of to lake concentrations, 257–262
  residual and cumulative error of compared, 284t–290t
  residuals, 252
  solute mass in, 259f–260f, 294t
solute flux, 331
  monthly water flow and, 121

relation to nutrient concentration and water flux, 110–111
seasonal climate and, 120–129
total annual, average proportion of, 115f
20-year volume-weighted average or median concentrations and ion balances for all sources of, 156t–157t
volume-weighted 20-year average or median concentrations for all sources of, 158t
solute imbalances, vs. water imbalances, 148f
solute input/loss, 256
solute mass and fluxes, 101–120, 331
average proportion of total annual solute fluxes, 115f
change in mass and chemical retention, 118–119
chemical budgets, 112–113
chemical input-output balance and chemical retention in the watersheds, 119–120
chemical retention in the lake, 115–117f
groundwater influxes, 109–110
in the lake, 101–109
precipitation influxes, 109
relation between water flux, concentration, and solute flux, 110–111
relation of inputs to outputs, 114
in the solute budget, 294t
solute outputs from the lake, 110
streamwater influxes, 109
time trends in annual input-output balance and lake storage changes, 102f–108f
solute mass input from tributaries, methods for determining, 237, 238f, 239
solute outputs, 110

solute retention, annual trends in, 117f
solute transport, 330
species composition, changes in, 318, 320
species diversity, 7, 9–10
in Cayuga and Mirror Lakes compared, 9
estimated number of, 15t
spectral pyranometer, 36
spill point elevation, 47, 58, 62
spring rains, 329
spruce (*Abies* sp.), 5
stable isotope ratios of oxygen method for calculating groundwater flux, 40, 214–216, 222t
statistical analysis system (SAS), 82
stilling well, 31, 32f, 209
stoneworts (*Nitella* sp.), 9
streamwater influx, 109, 149–150, 154, 307
streamwater inputs, daily, 81
streamwater outputs, 91, 93
sugar maple (*Acer saccharum*), 318
sulfate, 333
annual fluxes during driest/wettest years, 198t/199t
annual volume-weighted concentration and year, analysis of relation between, 160t
annual volume-weighted concentration, changes in, 317f
annual volume-weighted concentrations vs. normal distribution, 159t
from atmospheric deposition, 130, 152
average annual concentrations, 295t
budget, annual, 178t–179t
concentration and inputs of, 10, 84, 130, 131, 310f
decreasing concentrations of, 313, 331
groundwater inflow, 110, 114, 130, 242

sulfate *(continued)*
  groundwater outflow, 99–100, 110, 111
  long-term volume-weighted concentrations, 246
  mass, 101, 105f, 118, 196t, 331
  measurement of, 155t
  in Mirror Lake and Hubbard Brook Experimental Forest compared, 314f
  percent contribution to total influx/outflow in driest/wettest year, 200t
  and precipitation, correlation with, 331
  in precipitation, 130, 132
  relation between annual volume-weighted concentration and year for atmospheric precipitation, tributaries, and surface outlet, 161t
  relation between concentration and flow, 237
  retention, 116, 133, 142f, 144, 153, 242, 332
  seasonal fluxes, 121, 125, 149
  solute budget, 253, 255
  storage, variability in, 119
  in tributaries, 87, 91, 98, 114, 130, 231
  20-year average annual inputs, 162t–163t
  20-year average annual outputs, 164t–165t
  20-year volume-weighted average/median concentrations and ion balances, 156t
  volume-weighted 20-year average or median concentrations for, 158t
sulfur
  atmospheric deposition of, 312 *(see also* acid rain)
  pervasiveness of, 130, 152
  sedimentation rates, 3090f

sulfur dioxide (SO_2) emissions, 130, 134, 312
surface tributary mass flux, determining, 232
surface water, 31, 37–40
  contributions to chemical budgets, methods of determining, 278t–279t
  discharge, measurement of, 37, 39
  lake water, 233–235
  sample collection, 75–76, 231–235
  surface outflow, 232–233
  tributary inflow, 231–232
  uncertainties in, 208–211
surface water inflow (SWI), 31, 50, 56, 64
  tributary inflow, 208–209
  uncertainties in, 209–211
surface water outflow (SWO), 29, 50, 64, 232–233, 330
  back-calculation of, 210
  chemical mass loss by, 239–241
  discharge values, comparison of, 211
  measurement of, 209–211
  nutrient concentrations, 99–101, 232
  over the dam, 329
  sample collection, 232–233
"swimming rock," 2

temperature
  air, increases in, 5
  depth-time isopleths, 8f
  and spring streamflow peak, 63–64
temporal trends, 301, 319
terrestrial evapotranspiration, 28
thermal stratification, 7
thermal survey, 8f, 36–37
thermistor, 35
thermistor psychrometer, 35, 36
total dissolved solids, estimated concentrations of, 305f
total inputs, 140

total ionic charge, annual average over time, 86f
transpiration, 28, 302
tributary(ies), 2, 3f, 44, 231
 chemical mass input from, 236–239
 Northeast, 2, 4, 10, 28, 76, 78, 87, 98, 111, 119, 135–136, 138–139, 148, 153, 304, 332
 Northwest, 2, 28, 87, 98, 139, 153, 226, 304, 332
 West, 2, 28, 78, 87, 98, 139, 153, 304, 332
tributary inflow, 28, 109, 239, 329
 error in discharge, 209
 long-term average, 329
 mean absolute error in mass flux, determination of, 231
 measurement of, 208–209
 sample collection, 75, 231–232
 trends in, 94, 98–99
tributary outflow
 mass flux, determining, 232
 sample collection, 75
tributary runoff, Hubbard Brook Experimental Forest, 69
tributary water
 ammonium measurements, 78
 chemistry, 87–91
 daily fluxes, 81
 nutrient concentrations, annual volume-weighted average time trends, 88f–90f
20-year averages
 annual chemical inputs, 162t–163t
 annual solute outputs, 164t–165t

uncertainty, need for measurement of, 263
undersaturation, 7
Union of Concerned Scientists, 318
unmeasured cation, 134, 154
uptake, 330

vapor pressure, measuring, 207
vegetation, evolution of, 5
volumetric characteristics, 15t
volume-weighted average (VWA), nutrient concentrations, 80, 81, 88f–90f, 92f
volume-weighted concentrations
 annual of calcium, magnesium, potassium, sodium, chloride, hydrogen ion, and acid neutralizing capacity (ANC), changes in, 317f
 average or median chemical concentrations, 158t
 monthly average outlet data, 77
 and normal distribution compared, 159t
 and year, relations between, 160t

water
 annual fluxes during driest/wettest years, 198t–199t
 atmospheric, 33–37
 gains and losses of, 28
 ground (*see* groundwater)
 inputs and seasonal chemical patterns, 154
 loss of from Mirror Lake on annual basis, 29
 percent contribution to total influx/outflow in driest/wettest year, 200t
 samples, collection of, 73–74
 sources, 28
 surface (*see* surface water)
 20-year average annual inputs, 162t–163t
 20-year average annual outputs, 164t–165t
water budget, 23
 annual, components of, 51f
 atmospheric water, 33–37
 average annual, summary of, 194t–195t

water budget *(continued)*
  average lake, 44–46
  from bedrock sources, 27
  calculated from oxygen-isotope method, 40, 214–216, 219, 222t
  calculation of, 211–214
  and chemical budgets, uncertainties in determining, 226–229
  components averaged over time, 45f
  components, cumulative error in, 266f–271f
  components, mean monthly values of, 56f
  components, methods of determining, 31–44
  cumulative error, 58, 59f, 228
  error analysis, 43–44, 58
  evaporation, uncertainties in, 207–208
  groundwater, effects of, 71, 211–220
  hydrologic steady state, equation associated with, 215–216
  hydrologic variables, measurement of, 32f
  inter-annual variability, 45, 50–55
  mean monthly lake stage and mean monthly sums, 66f
  measurement accuracy, 43, 230
  monthly lake stage, 60f–62f
  month-to-month variability, 58–64
  precipitation, uncertainties in, 206–207
  relation of to lake concentrations, 257–262
  seasonal variability, 55–58
  surface water, 208–211
  20-year average of annual, 65f
  uncertainties in, 205, 331
  variability of, 50–64
  volume represented by residual of, 265f
  water storage in the lake, 31, 33, 205–206

water density, 7
water depth, 233
water-flow dynamics, 330
water flows, measuring, 205
water flux
  relation to nutrient concentration and solute flux, 110–111
  uncertainties, sources of, 225–226
water imbalances, solute imbalances vs., 148f
water level, declines in, 329
water lilies (*Nuphar* and *Nymphaea* spp.), 9
water lobelia (*Lobelia* sp.), 9
water loss
  sources of, 29–31, 57
  transpirational, 302
water quality, 4
watershed(s)
  changes to, 5
  chemical retention in, 119–120
  disturbances, 59
  Hubbard Brook Experimental Forest, 69
  input-output balance, 119–120
  maximum relief, 2
  nutrient export, 119
  outline of, 3f
  precipitation and, 28
  processes, solutes in, 314
  size, 28
  solute, retention trends, 142–143
  weathering within, 12, 119–120, 308
water sources, chemical characteristics of, 71, 83–101
water storage, 31, 33, 205–206
  annual estimates, effects of lake ice on, 233–234
  average annual change in, 45
  budget residuals of inputs, outputs, and change in, 276t–277t

changes in, 226–227, 240, 250
  errors in determining, 43–44
water table, 29, 32, 45
  average altitude of, 30f
  wells, 40, 76
water transparency, 7
water volume, 5, 205, 296t
weathering, 12, 119–120, 308
weir calculation, 39
wells
  chemical composition, variations in, 82
  index, 212, 235
  "slug tests," 43, 212, 214, 220
  stilling, 31, 32f, 209
  used to sample groundwater, locations of, 74f
  water table, 40, 76
West Tributary, 2, 28, 78, 87, 98, 139, 153, 304, 332
wet and dry conditions, biogeochemical influence of, 71, 149–154
wet-bulb thermistor, 35–36
white sucker (*Catostomus commersoni*), 9
wind, 7, 206

yellow perch (*Perca flavescens*), 9

zinc, 5
zooplankton, 18t

FRESHWATER ECOLOGY SERIES VOLUMES
WWW.UCPRESS.EDU/GO/FWE

Freshwater Mussel Ecology: A Multifactor Approach to Distribution and Abundance
  David L. Strayer

Mirror Lake: Interactions among Air, Land, and Water
  Thomas C. Winter and Gene E. Likens, eds.

Comparing Futures for the Sacramento-San Joaquin Delta
  Jay R. Lund, Ellen Hanak, William E. Fleenor, William A. Bennett, Richard E. Howitt, Jeffrey F. Mount, and Peter B. Moyle

Indexer:	Publication Services, Inc.
Composition:	Publication Services, Inc.
Text:	10.75/14 Bembo
Display:	Bembo
Printer and Binder:	Thomson-Shore